The Nature of Language

Dieter Hillert

The Nature of Language

Evolution, Paradigms and Circuits

Springer

Dieter Hillert
University of California, San Diego
School of Medicine
La Jolla
California
USA

ISBN 978-1-4939-0608-6 ISBN 978-1-4939-0609-3 (eBook)
DOI 10.1007/978-1-4939-0609-3
Springer New York Heidelberg Dordrecht London

Library of Congress Control Number: 2014936045

Printed on acid-free paper

Springer is part of Springer Science+Business Media (www.springer.com)

To my parents:

Charlotte Hillert, née Holland-Cunz
Guido H. J. Hillert

Preface

As an undergraduate student, I studied biology and philosophy at the University of Mainz and felt drawn to topics that relate cognitive phenomena to biological mechanisms. I felt as of today particularly inspired by the work of Derek Bickerton, Noam Chomsky, Charles Darwin, Hoimar von Ditfurth, Paul Feyerabend, Jerry Fodor, Eric Lenneberg, Karl Popper, and William van O. Quine.

After graduate studies at the Goethe University Frankfurt and RWTH Aachen University, I published my first book in German entitled Mental Representations of Word Meanings. Subsequently, I worked as post-doc at the Centre Paul Broca in Paris and EHESS and in Massachusetts at Boston University and MIT. Just before the reunification in Germany, I published my second German book Language Structures and Knowledge Representations. I resumed my work on the science of language at the University of Manchester, Science and Technology in England and at the University of California in San Diego. The autobiographical fragment serves here to acknowledge the institutions that provided support to my work.

The nature of language, aka the neurobiological foundations of language, plays a major role in the nature of language. The present book hopes to raise even more attention to this challenging but exciting interdisciplinary research area. Today, new empirical research comes out in large amounts faster than ever. Thus, I must admit that the selected topics are my subjective preference, and I am certain not to have addressed all research relevant to the questions and issues raised. Thus, I did not come close to an exhaustive survey of the literature referring to the nature of language. However, hopefully I presented sufficiently enough to illuminate how these interdisciplinary approaches in this field work and why it is actually a fascinating approach. Keeping this in mind let me introduce "The Nature of Language" with a modified aphorism by Hoimar von Ditfurth (1972, p. 245): *"We are, to put it in this way, in fact the *H. erectus of tomorrow."*[1]

San Diego, California — Dieter Hillert
December 2013

---

[1] In: H. v. Ditfurth (1972). *Im Anfang war der Wasserstoff* [German]. Hamburg: Hoffmann und Campe. The original quote used Neanderthals instead of H. erectus.

Introduction

The research area The Nature of Language is continuously growing integrating new methods and knowledge from different fields, and drives into new specialized subfields. The three parts dividing the chapters of this book are considered as significant thematic cornerstones. However, not all important topics can be addressed, but the selection may provide a starting point for further readings beyond the spectrum presented. Thereby, the thematic selection discusses some basic research questions from different angles. They include, but are not limited to:

- How did the human language system evolve?
- What are the neurogentic foundations for language?
- How can we map language processes to neural computations?
- How do we acquire and learn language(s)?

The first part Evolution presents evidence about the human linage and how it relates to the rise of language and cognition. Here, we consider that protomusic may have played a particular role in the evolution of speech and language. We assume that the ability to modulate vocalization has been the primary trigger for the evolution of language. Different stages are suggested from basic cognition to modern language, whereas our early ancestors—in particular H. erectus—might have used forms of communication still reflected in today's languages. In addition, some relevant biochemical mechanisms scaffolding the development, regulation, and maintenance of neural structures associated with language processing, are discussed. The caudate nucleus and basal ganglia, for instance, are significantly involved in speech and can be associated with forkhead-box P2 transcription factors, known as FOXP2. Comparative studies about the communicative behavior of non-human vocalizing species such as birds and whales provide further substantial details about the evolving vocalization mechanisms in different species and how the human language systems might have evolved. A possible evolutionary scenario will be described, which considers a gradual cognitive development from basic to complex communication systems.

The second part Paradigms introduces the concept of the biological disposition of human language. Typically, the language system operates left-sided within fronto-temporal circuits. Particular cortical regions as well as specific dorsal and ventral fiber tracks play a significant role in language processing. We discuss, moreover, our theoretical understanding of how the human language system might be structured. Different cognitive and linguistic approaches and models are presented, which make specific assumptions about the representations and computations of semantic, syntactic, lexical, and figurative information. Building a bridge between theoretical concepts of linguistic cognition and concepts of a neurobiological network or the unification of these approaches are challenging but at the same time intriguing. We introduce and discuss concepts such as natural semantics, binding theory, dependency grammar, artificial neural networks, lexical concepts and constructions, and universal semantic categories. Moreover, we cover the grammar of figurative speech and other idiosyncrasies, which seem to play a particular (or no) role in standard linguistic models.

The third part Circuits emphasizes the neural regions and circuits associated with sentence and/or lexical computations. Different types of electrophysiological (e.g., event-related potentials, magneto-encephalography) and neuroimaging evidence (e.g., structural and functional magnetic resonance imaging) evidence are presented. We discuss in this context the role of the left inferior frontal gyrus and verbal working memory functions in sentence processing and how different data might be accounted for by variance of sentence complexity and other context-dependent factors. Lexical concepts, however, are accessed broadly throughout the cortex. Some lexical concepts are closely associated with sensory-motor representations, others rely more on abstract, conceptual representations. Furthermore, we address the question how the brain computes figurative language as compared to literal language. Neuroimaging data on figurative language indicate that the recruitment of particular cortical regions depends on the linguistic structure of an expression (similar to literal language), but also on the integration of cross-domain knowledge. Particular portions of the parietal lobe, which are part of the mirror neuron system, may have played a significant role in the evolution of concept formation, consciousness, and language.

The final chapters are reserved for issues related to acquisition and (re)learning of one or more languages and how the languages system breaks down in context of a particular medical condition. Here, we discuss three different medical conditions: aphasia, Alzheimer's disease, and autism spectrum disorder. Aphasic syndromes or symptoms are mainly caused by a stroke. Spontaneous post-stroke recovery involves the reorganization of relevant neural circuits and typically the brain shows to some extent unexpected plasticity. In contrast, language degrades along with the progressive decline of cognitive abilities in mild cognitive impairment and Alzheimer's disease. Syntactic and lexical processes are affected as well as working memories functions. Neuroimaging studies let us assume that the brain tries to compensate for degraded processes by recruiting broader and more remote cortical regions. Finally, we discuss autism spectrum disorders, which affect the ability to mentalize and interact socially. Although autism spectrum disorders cannot be considered as a homogenous group, most subjects have in common that they show atypical behavior with respect to figurative and pragmatic aspects of language as well as in tasks involving the theory of mind. The neuroimaging evidence can be considered as inconclusive. A subgroup of children with autism shows an unusual brain growth, which presumably results in atypical connectivity and pruning. Again, neuroimaging data reveal degraded activations in various cortical regions including the prefrontal cortex. The attempt to draw a picture about how our language system works may help to understand and treat these and other neuropsychological conditions involving language and communicative disorders. In sum, the Nature of Language, which is subtitled Evolution, Paradigms, & Circuits, aims to shed light from a variety of different disciplines and approaches on all these questions, statements, and results. The chapters will hopefully inspire to drive and expand the avenue of this fascinating field.

Contents

About the Author

Dieter Hillert, born in Germany, is a cognitive scientist and best known for studying language through cognitive, neurobiological and comparative approaches. He holds positions as an Adjunct Professor and Research Scientist, and is affiliated with San Diego State University, University of California, San Diego, and University of Utah. He also works as a science writer on topics related to mind and brain.

Part I
Evolution

Chapter 1
The Human Lineage

1.1 An Overview

How did language evolve? To approach this obviously mysterious question, we would need to inquire about the evolutionary path of Homo (H.) sapiens—aka – modern humans. The brain of modern humans is equipped with a computational system that provides significant and superior cognitive power. A subsystem of this computational system is the linguistic system. No other biological organism than modern humans has mastered cognitive skills to express inner states, feelings, thoughts, and ideas or to communicate information by using a complex language system, including prosodic, lexical, semantic, and syntactic computations. We do not imply thereby that the ancestors of modern humans did not have language capacities. Instead, we state here, that the biological capacity scaffolding our linguistic system, gradually evolved over millions of years. Thus, for understanding the blueprint of the origin of language it is important to obtain a precise picture about the factors that support those neurobiological processes involved in communicative computations. We refer here with the term "language" to all natural languages, living or dead. The capacity to speak a language is based on universal computations of the human mind and this skill set enables us to create and express infinitely new meanings. What is finite about language are the sets of linguistic rules based on these universal parameters, which in turn are predetermined by species-specific neurobiological dispositions.

Modern humans share a common ancestry with other human primates. Millions of years divide between different species and presumably their evolution is the result of gradual genetic mutations determined by factors we can only speculate about at present. While language is the result of the genotype of H. sapiens, we cannot find evidence for the idea, as we discuss throughout, that this genotype is specifically designed for developing language. Modern humans are equipped with a biological disposition for language (BDL), but it does not result from a single, massive mutation. The human genotype supports specific cognitive properties, which are essential for the acquisition of language as well as for other cognitive capacities. Thus, here we point to a range of different assumptions about events that possibly triggered relatively small mutations in context of natural selection. Many factors

D. Hillert, *The Nature of Language*, DOI 10.1007/978-1-4939-0609-3_1,

over time may have shaped the neurobiological processes that allow us exchanging information about the world and how things work, but which can be also used for expressing internal states about emotions, opinions, beliefs, or attitudes. Most of all, language provides social stability among members of a group or population, and it is possible to discuss common goals and intentions.

An influential philosophical view of the twentieth century postulates that only modern humans are born with an innate linguistic universal grammar and implicitly rejects the assumption of a gradual evolution towards a BDL (Chomsky 1995; Bickerton 2009). To abandon the concept of language evolution is an irrational stance and ignores tremendous progress in life-sciences. This creationistic position was (and partly is) particularly popular as it is immune against any empirical data. Otherwise, it reflects the difficulty to develop a plausible cognitive model of primate evolution. For instance, while the age of a primate fossil can be approximately determined with radiometric or incremental readings, it is difficult to understand how these fossils relate to each other and thus how to classify them as part of an evolutionary tree. Even the isolation of mitochondrial (mt)DNA[1] in fossil bones is difficult as DNA degrades over time in dead tissues and bones. In general, it is assumed that fossils older than 40,000 years do not entail DNA. However, the BDL is a result of a long evolutionary process and our cognitive capacity is not as unique as it has been claimed to be in past: each communicative capacity is specific to a particular species, but their interpretations are often anthropomorphized and considered as inferior to the BDL. A better understanding of our linguistic capacities requires considering comparisons to other, non-human communication systems as the BDL is after all the result of natural selection and not of a metaphysical resolution. In this vein, Paul Broca stated once: *"I would rather be a transformed ape than a degenerate son of Adam"* (quoted by Sagan 1979). Before we discuss some possible scenarios that might have triggered and supported the evolution of cognition, let us first review want we know at present about the evolutionary path of mankind.

One of the first ideas about how to examine systematically the relations between single languages and their origin was to describe natural languages in analogy to the assumed evolution of species. August Schleicher (1861), a linguist, used a family tree model that resembled a botanical taxonomy to describe the history of languages in terms of developmental stages such as maturity or decline. At the same time, he was an advocate of a polygenetic account. That is, he assumed that several language groups developed from a speechless ape man (the "Urmensch"), who had ape-like ancestors. Schleicher believed that the large variety of languages, which developed independent from each other, speaks against the idea of a protolanguage, from which modern languages have been derived. In this vein, the zoologist and generalist Ernst Haeckel (1868) argued that different species and races with their different languages are descend from an ape man. Accordingly, many linguists of the late nineteenth and twentieth century believed in a direct relationship between a human race and

[1] mtDNA is in most species, including humans, inherited from the mother. It is located in mitochondria, structures that convert the chemical energy from food into a form that cells can use and can be regarded as the smallest chromosome.

the particular language spoken, mostly for the purpose to defend a view of different lineages. However, Alfredo Trombetti (1905), an Italian linguist, advocated the monogenetic account in line with Charles Darwin's (1871) viewpoint and assumed that all languages can be traced back to a language spoken 100,000–200,000 years ago. More recently, Joseph Greenberg (1963) worked on a comprehensive typological classification of all world languages with the goal to find linguistic universals. He shared his goal with Noam Chomsky (1956), whose approach is deductive and theory-driven rather than empirical. It remained equivocal how to link these findings to the neurobiological architecture of the human brain. For instance, some universal linguistic properties are considered as absolute such as that all language have pronouns; others express tendencies such as that languages with a canonical word-order (SVO, subject-verb-object) typically feature prepositions rather than postpositions, which can be found in SOV languages (e.g., in the phrase *He gave the book to her* the particle *to* is placed before the indirect object argument; in the Japanese version of this phrase *Kare wa kanojo ni hon o ageru* the particles or case markers are placed after the subject and after both grammatical objects).

1.2 Fossil Evidence

Today, evolutionists from different disciplines compare biology, cognition, and behavior of different species to understand the evolution of those properties that constitute the BDL as already pioneered by Charles Darwin (1871) and much later further developed by Eric Lenneberg (1967). However, speculations and beliefs have often their roots in insufficient empirical evidence and rely on our imagination of possible events, which occurred during an epoch of several thousands or even millions of years. One plausible approach is to focus on external clues. For instance, the parallel findings of particular tool and fossil discoveries, which can be associated with cultural-behavioral changes and signs of symbolic meanings, are indicators of more sophisticated forms of communication. In addition to the ability of verbal communication, modern human populations shared cultural behaviors such as crafting tools and cloth, fishing, bartering, decorating and self-ornamentation, creating symbolic art forms, playing games and music, and commemorating the dead. If the origin of the BDL coincides with behavioral modernity, an approximate timeline for the use of modern language may be around 50,000–40,000 years ago. However, behavioral modernity and more sophisticated forms of communication did not emerge out of nowhere and the origin of the BDL can be traced back to our closest living relatives and possibly beyond (see Fig. 1.1).

In addition, common chimpanzees (Pan troglodytes)[2] and genus Homo split in an epoch of about 7–4 million years ago (mya). While archaeological data sustain the account that humans and chimpanzees separated first 7.4–6.5 mya, DNA

---

[2] Chimpanzees are great apes and they belong to the genus Pan as do Bonobos, also called pygmy chimpanzees (Pan paniscus).

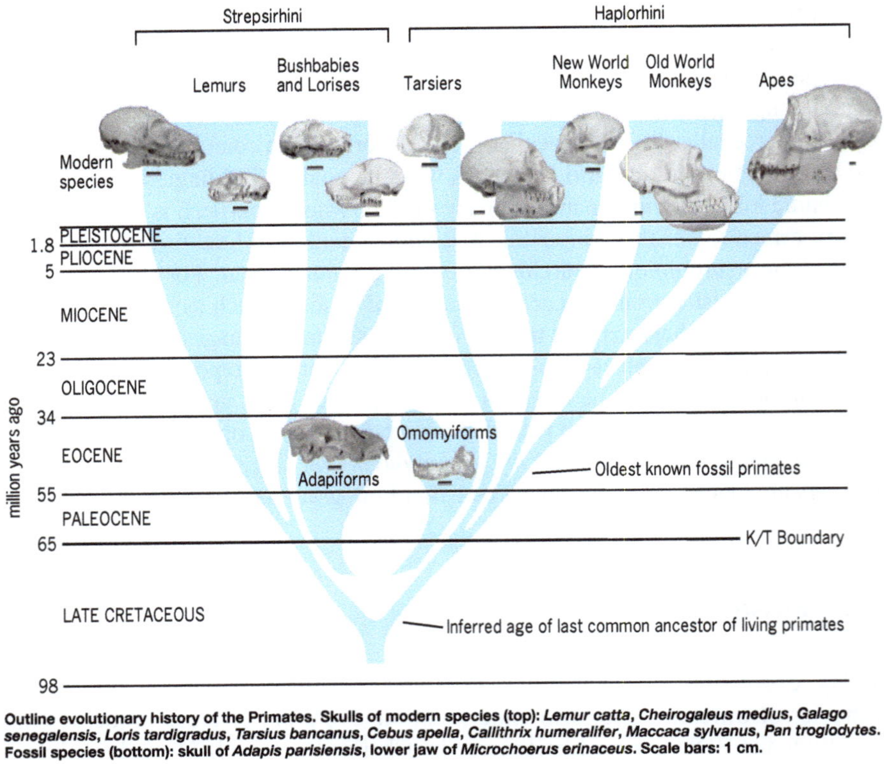

Fig. 1.1 Diagram of major evolutionary splits in the primate evolution. (Adapted Soligo's Website UCL, Anthropology)

analysis shows that the human lineage separated for good from the chimpanzees by ca. 4–5 mya. Interestingly, Sahelanthropus tchadensis, a 7 my old fossil with an estimated brain size of 320–380 cc, appears to be much more human-like than a 4–3 my old ancient human species (e.g., Wood 2002). The human-chimpanzee speciation took place over a long period of time and it is to assume that interbreeding took place before human ancestors definitely separated from the *Pan* lineage. This is a plausible assumption as our biological closest relatives, the common chimpanzee and the bonobo, are able to produce off-springs, although they split 1 mya. A direct comparison between different types of speciation is of limited value, but it points to the critique that the concept of species as developed by Ernst Mayr (1942) needs further refinements. Here, we note that some, yet unknown factors triggered the evolution of the BDL in the human lineage but not in the lineage of other great apes including the genus *Pan*. The geographical separation of hominids and "proto-chimpanzees" has been often considered as the reason for different lineages: The formation of the Rift Valley in East Africa, a dry savannah, might have isolated the evolution of hominids as chimpanzees would have lived in the wet jungles of Central and West Africa. The first chimpanzee fossil, three teeth dated 500,000 years old, was however found in East Africa, near Lake Baringo in Kenya, and not in Central and West Africa (McBrearhy and Jablonski 2005). Thus, we can doubt that the Rift Valley is a key factor for separate evolutionary paths of hominids and chimpanzee. Accordingly, the shift to the savannah may be also a weak hypothesis for the development of walking upright.

By comparing genes shared by modern humans and common chimpanzees, the Chimpanzee Sequencing and Analysis Consortium found that the FOXP2 (fork-head-box P2) evolved rapidly in the human lineage. FOXP2 is a transcription factor involved in speech and language. Chimpanzees are able to use vocalization, gestures and facial expressions for communication. However, this basic pre-disposition for language did further evolve by natural selection during the human lineage favoring the use of complex vocalization patterns. The evolution to modern humans occurred about 4 million years after the final split from chimpanzees. Anthropologists particularly focused on transitional forms, i.e., species with both ape- and human-like features to find evidence or signs for a particular classification scheme. Although the list of human evolution fossils is extremely short, such transitional species seemed to have been discovered. For instance, in 1974 about 40 % of the skeleton of an individual Australopithecus afarensis (AL 288–1)—nicknamed Lucy according to the title of Beatle song Lucy in the Sky with Diamonds—has been discovered in Ethiopia (Johanson and Edey 1981). Lucy was only 1.1 m (3'1") tall and her brain had an approximate size of 400–500 cc, which corresponds to 35 % of modern humans. However, apparently she walked upright as her pelvis and leg bones resembled those of modern humans. In contrast, the arm length of Lucy was considerably larger than of modern humans, probably favoring trees as habitat. This species dated 3.2 mya is certainly an example of physiological evolution towards bipedalism, but her brain size is more comparable to those of chimpanzees. Cranial capacity per se, however, does not need to correspond to cognitive capacities. Particular cognitive functions require specific neural wiring and brain size alone does not inform about the neural circuits available for communication. However, here we can rule out that Lucy had already a language-ready brain, but we cannot exclude the possibility that she used basic gestures and vocalizations to communicate with others of her kind.

Moreover, as Fig. 1.2 shows, an evolutionary significant change occurred in particular with the "birth" of the H. erectus at ca. 2 mya—an increase of 800 to 1400 cc in the Late Pleistocene, while the body size did not change significantly (Falk 2007).[3] But the diagram also shows that the cranial capacity continuously increased between the Australeoitheous group and late H. erectus group, if the groups Paranthropus and H. floresiensis (see fn. 5) are not considered. The evolution towards the capacity to process complex language structures requires, however, not only neurological rewiring and an increase of brain size, but also positional changes of the vocal tract connected to the lungs located in the chest. The need for vocalization may have indeed mutually triggered this neurological reorganization along

[3] The increase of the brain size resulted in some drawbacks. Human infants seem to be born premature when considering the typical correlation between brain size and gestation period. Compared to other primates, the human gestation period would be 17 months instead of 9 months. The brain growth in human infants slows down only one year after birth, that is, human gestation lasts 21 months.

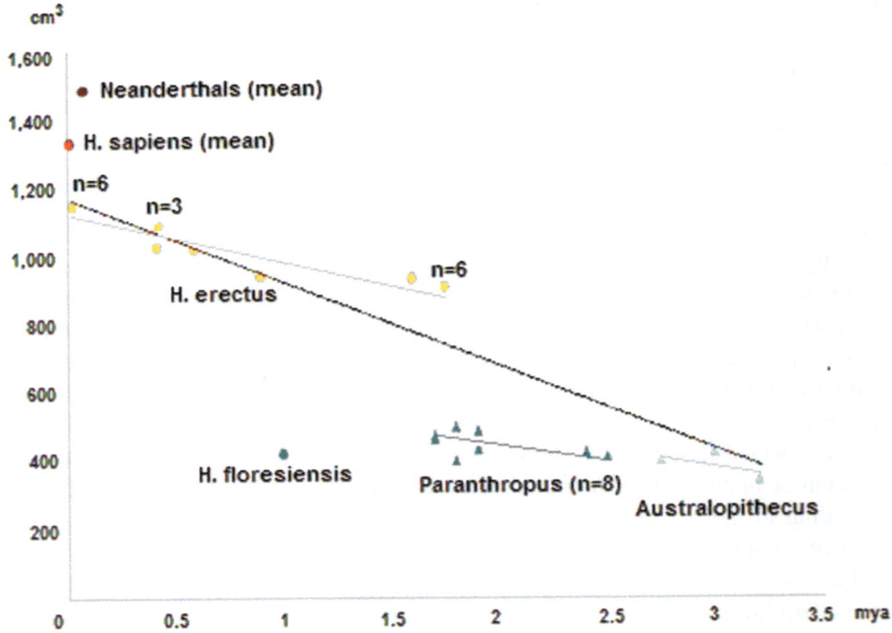

Fig. 1.2 Cranial capacities of adult hominins. If the Paranthropus group and H. floresiensis will not be considered, cranial capacity continuously increased in the human lineage. (Data Falk 2007)

with changes of the larynx position. However, brain rewiring is most important to allow complex cognitive processing. This adaptive reorganization may have been triggered by multiple factors, not only by intrinsic variable such as the need for vocalization, but also environmental conditions may have driven the cognitive and linguistic evolution in a particular direction. Let us return to the question, why did brain size significantly increase in the case of the H. erectus lineage? Different kinds of hypotheses are proposed. For instance, the climate change hypothesis states that sudden changes of the weather conditions and major climate changes forced our ancestors to plan ahead and prepare for significant environmental changes. Thus, an increase of neurocognitive wiring occurred to adapt to harsh and difficult living conditions. Another account, the ecology hypothesis, is to some extent associated with the climate change hypothesis. Since our ancestors migrated away from the equator, the ecological conditions required adaption to less food resources. Also, it has been assumed that with the migration toward north or south from the equator less pathogens needed to be fought off, which in turn would have boost brain development. The ecology hypothesis implies that if our immune system requires more energy in form of calories to combat parasites, brain development would have been neglected. Certainly, this is an interesting hypothesis, but it implies also that brain development has been suppressed before migration, an assumption which seems to be quite difficult to maintain.

A plausible hypothesis is that social competition influenced the dramatic increase in brain size (Geary 2002). Social competition may have triggered cognitive behavior to manage and organize smaller groups or even larger populations. The benefits of structured, intimate communities are that to some extent every member of such a community feeds its ego by profiting from shared resources and knowledge. Thus, cultural evolution parallels biological evolution and developed through multiple generations. Linguistic communication may have been the tip of the iceberg of social communication. Social and biological fitness are two sides of a coin and their co-existence increases the likelihood of competing successfully against predators and difficult environmental conditions. A relatively rapid increase of brain size as in the case of the H. erectus may have its roots in an increase of social complexity.

Closely related to this social complexity hypothesis is Dunbar's gossip account, which claims that language emerged from gathering news from other people or from third parties. Its function is therefore to form alliances and believe concepts in the truth of statements they cannot verify with their own eyes. Based on studies with animals that behave socially, Dunbar found a high correlation between the typical frontal lobe capacity of the individuals of a species and the maximum size of the group they live in. Accordingly, the number of relationships to other individuals of the human community seems to have its limits. Intimate communities consist rarely of more than 150, who know each other. Communication among individuals of larger communities such as neighborhoods, educational systems or corporations does not occur face-to-face, but is based on regulations and rankings and includes subgroups that manage the role of each individual of a global community. The significant increase of the neocortex occurred ca. 1.8 mya during the epoch of the *H. eretcus*. In chimpanzees the neocortex occupies 50 % of the brain, but in modern humans 80 %. The neocortex is involved in higher order cognitive functioning including emotions, self-consciousness, beliefs, language, music, planning, belief concepts and complex ideas.

Figure 1.3 provides a more differentiated overview of the human lineage. The apparently bipedal Ardipithecus group consists of two subspecies: Ar. ramidus and Ar. kadabba. However, members of this group had an even smaller brain (ca. 300–350 cc) than Lucy but are considered as the evolutionary precursor of modern humans. Moreover, within the Paranthropus group it is questioned whether robust australopithecines (Au. or P. robustus) should be considered as a descendent from gracile australopithecine (Au. garhi), which belongs to the genus Australopithecus. In general, however, the term "australopithecine" refers to both genera. More recently, a partial hominid foot skeleton was discovered in the afar region of Ethiopia (BRT-VP-2/73), which is dated to be 3.4 million years old (Haile-Selassie et al. 2012). The species differs from those of the A. afarensis (Lucy) but is more similar to the earlier Ar. ramidus. Thus, the co-existence of different species during the Pliocene epoch (P_0) indicates multiple bipedal adaptations. This fossil evidence shows that bipedalism exists before human species were able to use vocalization in a more elaborated form or before tools were used or manufactured. However, it cannot be ruled out, as mentioned before that some hominids already used a relative basic communication system, but which significantly differed from those communicative forms used by

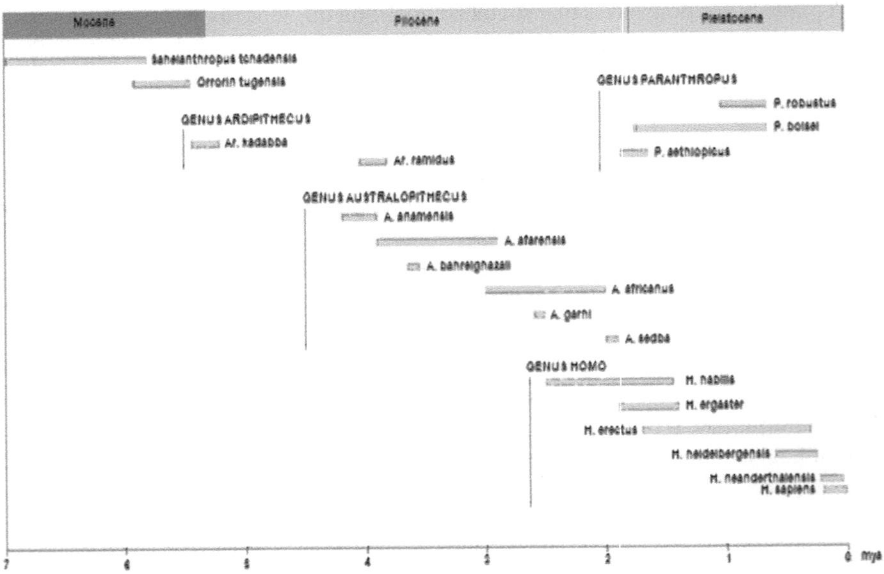

Fig. 1.3 Reconstructed ancestry of modern humans based on fossil evidence. (Adapted and modified, Wikipedia)

present-day common chimpanzees. The human lineage includes in the Paleolithic era (Old Stone Age) H. habilis (2.4–1.44 mya), H. erectus (1.9–0.3 mya), H. antecessor (800–500 kya; k = 1,000), H. heidelbergensis (600–250 kya), H. neanderthalensis (250–30 kya), H. rhodesiensis (300–125 kya), H. floresiensis (100–10 kya)[4], and H. sapiens (250 kya).

The location of the fossil findings and their age provides some clues about the possible migration patterns of different hominids and thus about the possible co-existence of different species during a particular epoch. Fossil evidence is quite rare and represent only a small percentage of the species ever lived on Earth. Thus, it seems to be plausibly that the reconstructed ancestry of modern humans reflects

[4] It is still today unclear whether the species "hobbit" (H. floresiensis), which had a brain volume of ca. 400 cc, should be regarded as a new species or a case of pathology such as microcephaly (e.g., Brown et al., 2004; Falk et al. 2005; Holloway et al. 2006). Although it appears as a form of microcephaly, the small hobbit brain seems not to fall in the category of microcephaly as defined today. This is supported by the fact that their body mass index (BMI) is comparable to modern humans. Also, the comparison of the LB1 endocast with great apes, H. erectus, and Australopithecus, modern humans, pygmy and microcephalic modern human indicates that LB1's brain shape resembles that of H. erectus (expanded frontal and temporal lobe), although with respect to the brain size it is more comparable to an Australopithecus. H. floresiensis seems therefore to be a separate species, which is closely related to H. erectus. Although the brain size of LB1 was small, the cortical structure was presumably advanced.

only a small fraction of hominid species that lived in the time span of the last 4 million years. But how did hominids migrated across the globe?

In the past, two main opposing hypotheses were discussed: the multi-regional hypothesis and the recent single-origin hypothesis or better known as the recent Out-of-Africa hypothesis.[5] Both accounts assume that the birth of human genetic diversity came out of Africa, but the multi-regional hypothesis makes strong claims about hybridization to account for regional continuity (Wolpoff et al. 2000). The assumption is that the human species appeared in the beginning of the recent period of repeated glaciations, the geological epoch Pleistocene (ca. 2.5 mya–11,700 years ago). Subsequent evolution took place in different geographic regions from H. erectus to modern humans, while there was a lateral gene flow between these different human populations. Thus, in every region some local adaptions were maintained along with general properties common to all regions. Wolpoff et al. (2001) refer to character traits of modern human's skull fossils in Australia and Central Europe and assume a separate ancestry from Java H. erectus for Australia and from Neanderthals for Central Europe. However, the opposing recent single-origin hypothesis states that modern humans evolved as a new species in Africa ca. 100–200 kya and dispersed from Africa ca. 50–60 kya to replace already existing human species in the new regions without hybridization (Weaver and Roseman 2008).

The debate about how the hominid migration took place is far from closed. Accordingly, Trinkhaus (2007) provided an analysis of numerous fossil features (e.g., aspects of the skull and mandible shape, shape and size of tooth and other bones) of later European humans that could not be found in African samples but in the Neanderthal sample. He assumes a "modest level of assimilation of Neanderthals into early modern human populations as the latter dispersed across Europe" (lateral genetic flow). While early humans are silent witnesses as they testify through bones and tools, the female-specific mtDNA and the male-specific Y-chromosome of modern humans carry a familial signature. Random mutations in both cases allow determining the degree of kinship and ancestral origin. DNA evidence shows, of course, that all humans in this world are very closely related compared to, let say to chimpanzees from different African groups. The analysis of mtDNA revealed that Mitochondrial Eve, also called African Eve, is the most recent common ancestor (MRCA), a woman, who lived about 200 kya in Africa, from whom all living humans today are descend (Cann et al. 1987). Y-chromosome data also support the single-origin hypothesis and indicate an expansion back from Asia to Africa (Hammer et al. 1998). Moreover, recent ancient Neanderthal DNA analysis seems to have overcome some analysis problems (contamination with modern human DNA) and revealed that Neanderthals share 1–4 % more genetic variants with Non-Africans than with the two Sub-Saharan African populations—the San from Southern Africa and the Yoruba from Western Africa (Green et al. 2010). Similarly, an mtDNA analysis shows that the human genus Denisova recently discovered in the Russian Denisova Cave is distinct from Neanderthals and modern humans, but lived at the same time ca. 41,000 years ago (Reich et al. 2010). Further analyses

[5] Other terms used for the recent single-origin hypothesis are "Recent African Origin model" and "Replacement Hypothesis."

show a 4–6% common genome with living Melanesian and Australian Aborigines people in Oceania, but with no other human population. Thus, lateral gene flow took place between two regions outside of Africa (Reich et al. 2011).[6]

In sum, the evolutionary path of modern humans is viewed at present as follows: Based on molecular clock computations, humans and chimpanzees diverged for good around 4 mya. Several models are discussed about when and how the split between chimpanzees and humans evolved. As mentioned above, genome analyses indicate that hominids and chimpanzees diverged first, but interbred later (e.g., Bower 2006). Their genes differ just by ca. 1.6%, a difference, which is smaller than between chimpanzees and gorillas (2.3%).[7] Genetically humans can be thus considered as a third or sister species of (common) chimpanzees and bonobos. Argon dating suggests that bipedal walking was developed to some extent by 4 mya, the oldest stone tools dating back to 2.6 million years, and the H. erectus dispersed out of Africa by ca. 1.8 mya, first to modern-day Israel, then to Asia and Europe. The first fossil of H. erectus was found in 1891 on Java by the Dutch physician Eugene Dubois, and he assumed that it was a transitional species between apes and modern humans. Another often quoted H. erectus fossil was found near Peking (Peking Man) and has been dated between 680,000–780,000 years ago (Shen et al. 2009). Today, it is questioned whether the early phase (1.8–1.24 mya) belongs to a different species, H. ergaster. Some anthropologists assume that H. ergaster is the direct African ancestor of H. erectus and migrated to Asia.[8] There is no doubt that this species walked truly upright as evidenced by locking knees and a different location of the foramen magnum.[9] Ancestors of anatomically modern humans (AMH) seem to have resided first in the East African Rift Valley, from which they migrated to other directions—north, west and south. It is assumed that one lineage evolved to Neanderthals in today's Europe and Middle East about 700,000 years ago and another one evolved to modern humans in Africa. Because of the end of the ice age about 100,000 years ago, the climate conditions improved in Africa and presumably led to a population growth of early humans. The oldest fossils of direct modern humans' ancestors are skulls from two adult males and a child found near the village

[6] The rate of DNA decay is largely temperature dependent. However, recent calculations indicate that DNA can be longer preserved than previously assumed. Allentoft et al. (2012) reported that frozen DNA (−5°C) has a half-time of up to 158,000 years, i.e., it would last ca. 6.8 million years. At ca. 13°C, they found a half-time of 521 years for Moa bones, 400 times longer than lab tests predict. (Moa is an extinct wingless bird, which lived in New Zealand up to 1400 AD.)

[7] Of course, we compare today's chimpanzees (including genes and cranial capacity) with hominid species, but not with those chimpanzees that split from the human lineage about 4 mya. It is implied that the evolution of the chimpanzee lineage was relatively limited although this conclusion may be premature.

[8] A wide range of fossil findings are comparable to the discovery of the Dutch anthropologist Eugène Dubois' in 1891 on Java, who named the fragment of a skull "Pithecanthropus erectus" (upright ape-man). While there is today no doubt that the Java man belongs to the genus Homo, today anthropologists prefer to use the term "Homo erectus" exclusively for hominids found on Java. Depending on the regions of the fossils discovery, terms such as Homo georgicus (Dmanisi, Georgia), H. soloensins (Ngandong, Java), H. pekinensis (Peking/Beijing, China) are used. The African variant of the H. erectus is typically called H. ergaster.

[9] The "foramen magnum" refers to the hole in the skull, where the spine enters.

Herto in Ethopia and dated around 160,000 years ago. The Herto skull is slightly larger (1,450 cc) than the average volume of modern human skulls (1,350–1,400 cc) and therefore there are considered as a separate subspecies: H. sapiens idaltu. AMH lived before most Neanderthals and thus they cannot be decedents from them. Interestingly, H. sapiens idaltu had a complex stone technology as more than several hundred stone tools were found in the same sediments. Further migration took place between 180,000–90,000 years ago. AMH quickly spread across Eurasia and seemed to have replaced other hominids. It is estimated that they reached China by 68,000 years, Australia 60,000 years, Europe 36,000 years ago and America and Oceania by 12,000 years ago (Cann et al. 1987). By about 10,000 years ago, modern humans migrated to all parts of the world with except of Antarctica and some islands (e.g., New Zealand and Hawaii).

Determining the origin of language among the human linage seems to be impossible in considering the small amount of evidence currently available. The use of spoken language may coincide with the use of cultural forms expressing abstract and symbolic meanings. We will explore below possible scenarios or signature events how modern humans developed language and how preconditions of the BDL may have evolved in ancestral human species such as Lucy. Despite of many open questions about the origin of the BDL, in general there is certainly no doubt that a complex communication system generates a significant evolutionary edge to survive against competing hominid species and other predators. An elaborated natural communication system scaffolds the development of cultural forms, it allows to exchange instant messages, express common goals, expectations and plans for the future, and most of all it enhances the social bond among members of a group or tribe.

Arnold Schleicher and Ernst Haeckel's assumed that different languages correspond to different species or races. This was a common view in the eighteenth and nineteenth century. Today, we acknowledge, that linguistic typologies as well as ethnic or character traits are unrelated to the BDL. Here, we try to increase our knowledge about the underlying universal cognitive and neurobiological principles that allow acquiring one or more languages. One of the prime questions is how did the BDL evolve? Did ancestors of modern humans already possessed a pre-condition for language and what were the factors for cognitive-linguistic adaptations and mutations scaffolding spoken language processing? Our approach considers the evolution of the BDL not as an isolated process, but which mutually evolved with other cognitive capacities characteristic for modern humans.

References

Allentoft, M. E., Collins, M., Harker, D., Haile, J., Oskam, C. L., Hale, M. L., Campos, P. F., Samaniego, J. A., Gilbert, M. T., Willerslev, E., Zhang, G., Scofield, R. P., Holdaway, R. N., & Bunce, M. (2012). The half-life of DNA in bone: measuring decay kinetics in 158 dated fossils. *Proceedings of the Royal Society B: Biological Sciences, 279*(1748), 4724–4733.

Bickerton, D. (2009). *Adam's tongue*. New York: Hill and Wang.

Bower, B. (2006). Hybrid-driven evolution: Genomes show complexity of human-chimp split. *Science News, 169*(20), 308–309.

Brown, P., Sutikna, T., Morwood, M. J., Soejono, R. P., Jatmiko, W. (2004). A new small-bodied hominid from the late Pleistocene of Flores, Indonesia. *Nature, 441,* 624–628.
Cann, R. L., Stoneking, M., & Wilson, A. C. (1987). Mitochondrial DNA and human evolution. *Nature, 325*(6099), 31–36.
Chomsky, N. (1956). Three models for the description of language. *Information Theory, IRE Trans, 2*(3), 113–124.
Chomsky, N. (1995). *The minimalist program*. Cambridge: MIT Press.
Darwin, C. (1871). *The descent of man, and selection in relation to sex.* London: John Murray.
Falk, D. (2007). Evolution of the primate brain. In W. Henke & I. Tattersall (Eds.), *Handbook of palaeoanthropology, vol. 2: Primate evolution and human origins* (pp. 1133–1162). Berlin: Springer.
Falk, D., Hildebolt, C., Smith, K., Morwood, M. J., Sutikna, T., Brown, P., Jatmiko, Saptomo, E. W., Brunsden, B., & Prior, F. (2005). The brain of LB, Homo floresiensis. *Science, 308*(5719), 242–245.
Geary, D. (2002). Principles of evolutionary educational psychology. *Learning and Individual Differences, 12,* 317–345.
Greenberg, J. (1963). *Universals of language*. Cambridge: MIT Press.
Green, R. E., et. al. (2010). A Draft Sequence of the Neandertal Genome. *Science, 328* (5979), 710–722.
Haeckel, E. (1868). *The history of creation.* London: Kegan Paul, Trench & Co.
Haile-Selassie, Y., Saylor, B. Z., Deino, A., Levin, N. E., Alene, M., & Latimer, B. M. (2012). A new hominin foot from Ethiopia shows multiple Pliocene bipedal adaptations. *Nature, 483,* 565–569.
Hammer, M. F., Karafet, T., Rasanayagam, A., Wood, E. T., Altheide, T. K., Jenkins, T., Griffiths, R. C., Templeton, A. R., & Zegura, S. L. (1998). Out of Africa and back again: Nested cladistic analysis of human y chromosome variation. *Molecular Biology and Evolution, 15*(4), 427–441.
Holloway, R. L., Brown, P., Schoenemann, P. T., & Monge, J. (2006). The brain endocast of *Homo floresiensis*: microcephaly and other issues. *American Journal of Physical Anthropology, 129*(42), 105.
Johanson, D., & Edey, M. (1981). *Lucy, the beginnings of humankind.* Granada: St Albans.
Lenneberg, E. H. (1967). *Biological foundations of language*. New York: Wiley.
Mayr, E. (1942). *Systematics and the origin of species*. New York: Columbia University Press.
McBrearty, S., & Jablonski, N. G. (2005). First fossil chimpanzee. *Nature, 437,* 105–108.
Reich, D., et al. (2010). Genetic history of an archaic hominin group from Denisova Cave in Siberia. *Nature, 468*(7327), 1053–1060.
Reich, D., et al. (2011). Denisova admixture and the first modern human dispersals into Southeast Asia and Oceania. *American Journal of Human Genetics, 89*(4), 516–528.
Sagan, C. (1979). *Broca's brain.* New York: Random House.
Schleicher, A. (1861). *Compendium der vergleichenden Grammatik der indogermanischen Sprachen* (German). Weimar: H. Bŏhlau. English version: Schleicher (1874). A Compendium of the comparative grammar of the Indo-European, Sanskrit, Greek, and Latin languages (translated an abridged version from the 3rd German edition by Herbert Bendall). London: Trübner and Co.
Shen, G., Gao, X., Gao, B., & Granger, D. (2009). Age of Zhoukoudian Homo erectus determined with (26)Al/(10)Be burial dating. *Nature, 458*(7235), 198–200.
Trinkaus, E. (2007). European early modern humans and the fate of the Neanderthals. *Proceedings National Academy of Science, 104*(18), 7367–7372.
Trombetti, A. (1905). *L'unit d'origine del linguaggio*. Bologna: Luigi Beltrami. (Italian)
Weaver, T. D., & Roseman, C. C. (2008). New developments in the genetic evidence for modern human origins. *Evolutionary Anthropology: Issues, News, and Reviews, 17*(1), 69–80.
Wolpoff, M. H., Hawks, J., & Caspari, R. (2000). Multiregional, not multiple origins. *American Journal of Physiological Anthropology, 112*(1), 129–136.
Wolpoff, M. H., Hawks, J., David, W., Frayer, D. W., & Hunley, K. (2001). Modern Human Ancestry at the Peripheries: A Test of the Replacement Theory. *Science, 291*(5502), 293–297.
Wood, B. (2002). Hominid revelations from Chad. *Nature, 418,* 133–135.

Chapter 2
Protomusic and Speech

2.1 The Role of Protomusic

Darwin (1859) did not discuss in his first great work "Origin of Species" evolutionary approaches to the human mind and thus avoided to argue against anti-evolutionist positions such as explicitly expressed by the linguist Friedrich Max Müller (1866). According to Müller, it is language, which is the key feature that divides man from beast. Darwin introduced the idea of a musical protolanguage, a concept that still represents an important aspect of modern debates. Comparable ideas were already expressed in the seventienth century, to which Darwin also refers. For example, it has been argued that bird songs have some function arising from male rivalry and territorial competition (Thomas 1995). Thus, although he took a multi-component approach towards language evolution, Darwin emphasizes the significance of complex vocalization by embedding his view in a broad theory of evolution (e.g., Fitch 2000; Egnor and Hauser 2004). Darwin (1871, pp. 880) mentions in his second great work "The descent of man and selection in relation to sex" some issues related to the evolution of language: "*... it appears probable that the progenitors of man, either the males or females or both sexes, before acquiring the power of expressing their mutual love in articulate language, endeavoured to charm each other with musical notes and rhythm.*"

In particular, evolutionists and their opponents agreed that emotions and memory are shared between humans and animals. Darwin's approach can be summarized in short as follows: First, he points out that the "language instinct" is not a true instinct but involves a learning process. By using the term "instinctive tendency to acquire an art", he does not solely refer to a language instinct but to an instinct of any form of more complex culture and cognition. Second, Darwin already recognized that the seat of the language capacity is the human brain while the human vocal tract alone is insufficient to empower humans with language. Third, he draws a parallel between human and bird vocalization: Birds have an instinct to sing, humans have an instinct to speak (in addition to singing) and the acquisition of regional dialects is both characteristic for bird songs and human speech (see Fitch 2013). Not only these general observations can be considered still as quite contemporary as they

D. Hillert, *The Nature of Language*, DOI 10.1007/978-1-4939-0609-3_2,

are even today of much debate but also his account of "musical protolanguage" to which we refer here as "protomusic."

First, Darwin assumes that the development of cognition and social intelligence represents the foundation for the rise of protomusic and then to protolanguage (Hewes 1973; Bickerton 1990). Some salient properties of protomusic would have been driven by sexual selection, and it has been used in courtship, territorial behavior and in expressing emotions such as love, jealousy, and triumph. Darwin emphasizes in particular that vocal imitation is a cognitive function, songbirds and humans are sharing. Initial communication might have started with vocal imitations and by associating vocalizations with objects, states, events, actions or emotions, and/or properties of thereof to create meanings (Holloway 1969, 1992). Among others, the imitations may refer to onomatopoeia (e.g., imitating animal sounds, sounds of natural events or sounds expressing emotions) or even to the instinct calls of our direct ancestors supported by gestures and signs. Darwin assumed that vocalizations have been the primary trigger for the evolution of language but not gestural signals as it would have been more recently advocated (e.g., Corballis 2003; Arbib 2005; Call and Tomasello 2007). Moreover, Darwin makes the plausible assumption that along with increased vocalization the vocal tract became gradually refined allowing a broader range of discrete utterances. His approach is highly plausible as Darwin refers to comparative data: vocal learning evolved in different clades in mammals such as bats, cetaceans, pinipeds, and in birds such as oscine songbirds, hummingbirds, parrots without referring to symbolic meanings. The precise development of symbols and propositions expressing meanings at the sentence level remains highly speculative. Here, however, more complex cognition certainly played an important role in further developments of communicative skills in our early ancestors. Thereby, it is to assume that the development of social competence and strategic planning including tool use contributed to the rise of more complex language competence.

Accordingly, it remains a secret how the concept of phrases or sentences was developed either out of single words or directly out of protomusic or a combination thereof. This question is reflected in today's debate about holistic protolanguage. The linguist Otto Jesperson (1922) claimed a holistic starting point as such that propositional meanings were derived from sung phrases. (Multi-)syllables of music phrases would have been isolated and associated with individual lexical meanings referring to different syntactic categories (parts of speech). This holistic approach implying synthesis by analysis was defended in recent discussions on the origin of language (Wray 1998; Arbib 2005) but extensively criticized by accounts favoring an analysis by synthesis approach, in which syntactic operations are learned to combine single words (Bickerton 2007; Tallerman 2007). Often certain models represent extreme scenarios and a more realistic picture is provided by using plausible properties of different accounts to approach the relevant scenario. Indeed, in assuming protomusic as a starting point, an interesting hypothesis is that global or holistic meanings were expressed by vocal sequences referring to complex events. The gradual process of fractioning resulted over time in discrete tonal units and/or phrases. However, before discussing in more detail a possible scenario about the origin of proto-communication and its evolving characteristics towards language, here we like to provide evidence for the assumption that vocalization *per se* was the starting point for the development of complex communication among humans.

2.2 Evolutionary Milestones

The primary form of communication among our closest relatives, the apes, is vocalization, whereas gestures or other non-verbal signals seem to have a supportive function. As mentioned before, other species prefer vocalization including ultrasound communication: dolphins, rodents, insects, avian, amphibian, and fish. Thus, vocalization is the primary form of social communication in a large number of different species. Thereby, initial stages of vocalization, the babbling stage, shows surprising parallels in humans, songbirds and other animals. Human infants babble spontaneously in the first year followed by different stages of language acquisition. Similarly, juvenile songbirds vocally babble by producing immature songs. For instance, baby zebra finches learn singing by listening to their tutor, typically the father. About 1 month after hatching, they babble a rapid stream of screechy tones and practice for hours. After 3 months of learning, the juvenile can exactly imitate the tutor's song. There are astonishing parallels of vocal learning and production between different species and further comparative research may inform us in more detail about the phylogenetic development of those (sub-)cortical systems supporting complex vocalization and therefore spoken language. The origin of vocalization in the history of human evolution, however, remains highly speculative, but we discuss in the following some possible scenarios without denying that alternative viewpoints are possible. Let us summarize in the following some significant milestones assumed to be involved in the evolutionary origin of modern language and which seemed to have built on each other:

Protostages

- Cognition: multi-sensory experience, instinct calls
- Music: rhythm, prosody, syllables
- Phrases: embodied concepts, basic phrases
- Speech: discrete sounds, symbolic concepts, basic syntax

The gradual increase of the cranial capacity and thus the rise of cognition might have already started as early as with the appearance of the Australopithecus ca. 4 million years ago (mya). It is believed that in the eastern part of Africa one of the australopith species, Aus. sediba, evolved about 2 mya into genus Homo. Thus, we claim at this point that our pre-human ancestors were generalist compared to the evolving human species. They might have used a variety of different communicative signals across different sensory-motor modalities including instinct calls, gestures, and grimaces to inform about dangers, benefits, or to express emotions. The multi-sensory interaction of various sources such as tones, noise, light, smell, tactile stimuli may have led to improved sensory-specific working memory (WM) functions, associative connections, and conceptual representations. Climate changes and/or the attempt to discover new ecological niches in the savannah may have led to more complex cognitive processing and thus favoring the evolution of a larger cranial capacity.[1] Thus, we assume that multi-sensory integration motivated by climate changes and social planning is the primary factor for the increase of cortical structures. Further specializations were probably triggered by various factors, in particular by the emergence of improved means of communication.

The use of protomusic, that is, of melodic-rhythmic patterns and harmonic passages, might be triggered by sexual selection, as proposed by Darwin (1871), parent care and group cohesion. Sexual selection expressed in courtship is an elaborative trait, which is beneficial in the competition for mates (Miller 2000). Also, motherese—the use of melodic and rhythmic utterances for infant offspring—is an important mean to provide infant care. Finally, protomusic might have been again beneficial for promoting social cohesion within groups (Brown 2000). Although, we argue here for an adaptationist approach, it should be pointed out that there are numerous alternative views including those arguing in favor of a non-adaptationist view of music evolution. According to a non-adaptationist stance, natural selection did not drive the origin of music. A variety of claims were made arguing, for example, that music grew out of impassioned speech (Spencer 1857) or incidental peculiarity of the human brain (James 1890; see also Pinker 1997). However, according to an evolutionistic neurobiological stance, it remains a plausible hypothesis that modern music as well as modern language are the product of natural selection evolving from protomusical behavior.

While we certainly can find arguments to explain the important role of protomusical behavior, it is difficult to model a gradual specification towards language. Thereby, the evolution of gradual changes is not disputed, but questions about the cultural–behavioral influence on genes and vice versa. We assume that our distant ancestors use their vocal tract to produce a variety of different pitches and volumes, perhaps accompanied by dance steps. Moreover, it is difficult to speculate where the concept of harmony and melody comes from but the concept of timing seems to be here fundamental. Timing involves the coordination of events, to synchronize processes and is biologically rooted. Life is timing and each organism has an inner clock, which must operate according to specific sequences. This concerns every aspect of an organism from heart beats, action potentials to speech coordination. Our distant ancestors must have discovered rhythm and harmony, perhaps by listening to their own heart beat and by means of ritualized behavior. The innate capacity to perceive harmonic patterns and melodies may indeed be a reflection or an expression of the principles of mother's nature that creates life. These biological principles may have represented the foundation for mapping biological structures to cognitive behavior. We further assume that cognitive specialization took place in various domains, which resulted in a cognitive capacity today known as human intelligence.

[1] Some recent studies indicate that our distant ancestors were not forced to leave rain forest in northeast Africa because the savannah expansion replacing forest territories. In fact, it is unlikely that at any time in the last 12 million years forest was extensive in the northeast Africa (Feakins et al. 2013). Thus, it appears that our ancestors tried to adapt to new ecological niches.

Modern language can be considered as one product of this cognitive evolution. But let us further speculate how modern language evolved from protomusic.

The next stage we postulate here refers to protophrases. Here, protophrases are vocal sequences derived from protomusic, whereas prosody and intonation has been at least partly preserved. It contrast to the melodies of singing, however, protophrases represent discrete sequences or icons used to refer to objects. Our distant ancestors may have vocalized to imitate other animals and environmental sounds and/or they produced random syllabic structures to express feelings. Thus, here we assume that phrasal or syntactic patterns were already used before the creation of symbolic concepts. These protophrases may have consisted of one or more units and were iconic, synthetic, and non-holophrastic in nature. Thus, we assume in line with Bickerton (2003) that it is unnecessary to postulate phonologically complex holophrases. The proposed phrases may have included also expressions what we call today proper nouns. Even apes can differentiate between different members in a group and have an idea about kinship relations. Primarily sensory embodied, single objects may have been named.

Because of lack of evidence, dating these different stages is not possible, but we speculate here that the discovery and development of symbolic behavior may correlate with a significant increase of cortical volume. In particular, the Homo erectus/Homo ergaster, who had an increased brain size of up to 1,100 cc, is a candidate for this development as this hominid species seems to have used more elaborated social structures, much like modern humans, and used sophisticated tools occasionally found with their fossils. Thus, we postulate that protophrases were used in the next stage as discrete units in a symbolic fashion. Thereby, units referring to conceptual categories form conceptual cluster among themselves. These clusters are today known as syntactic structures, which were gradually used in a recursive fashion to form subclauses. The development of discrete units at the basic lexical and syntactic level is perhaps the stage, which is typically referred to as protolanguage. Here, we use the term protospeech as human communication was primarily vocal by using vowel-consonant clusters. It is a stage, at which language has been born, a new cognitive ability with musical properties as residuals. The remaining development towards modern language includes everything what modern languages are made of. In addition to more complex phonological and syntactic structures, constraint by memory limitations, a large conceptual lexicon was acquired and taught from generation to generation. The evolution of the language system, which is even today an ongoing process, may have reached the modern stage not before the appearance of the Homo sapiens. Gradually specific linguistic features were added as they turned out to be useful for expressing specific meanings. For instance, the use of closed-class elements, words, which carry meaning only in a sentence context (e.g., determiners, pronouns, conjunctions, particles), may be a product of a late stage as it may be the use of inflection and derivations or verb argument structures. Today, we know that all (possible) human languages conform to a certain set of parameters, which presumably can be mapped onto innate properties. There is a dispute about how strict the set of parameters should be, but there is no doubt that biological constraints determine the acquisition and use of modern languages in H. sapiens.

Label	Function	Structure	Timeline (mya)
Cognition	Multi-modality	Multi-sensory experience, instinct calls	4
Protomusic	Sound imitation	Rhythm, prosody, syllables	Aus. sediba
Protophrases	Iconic reference	Embodied concepts, basic phrases	2
Protospeech	Symbolic reference	Abstract concepts, phonemes	H. erectus
Language	Sentences	Complex grammatical dependencies **Figurative language**	0.2 H. sapiens

Table 2.1 Possible milestones in the evolution of modern language. Function and structure built on each other. Here it is assumed that protomusic represents the foundation for the development of language and music in H. sapiens. Other non-verbal routines such as gestures or dance are considered to supplement vocalization and articulation.

Thus, here we ask which hominid species (closer than chimpanzees) had which property of modern language as it is a product of a continuous biological and cultural evolution rather the result of a sudden mutation. Our descriptions of the different stages can be regarded as a preliminary approximation for drawing a plausible picture of the evolving human language capacity. The human language faculty is not only the result of a cultural process that modifies lexical and grammatical knowledge across many generations, but it is also a biologically adapted cognitive capacity that evolved over millions of years. Table 2.1 details the proposed evolutionary milestones described above. Certainly, we would not be surprised if these ideas attract criticism, but taking into account the spare evidence available, the selected approach seems to be at present quite plausible.

The evolution of cognition is based on processing and integrating sensory-motor experiences. The occipital-temporal-parietal connections of the cerebral cortex integrate these data. In humans, in particular Wernicke's area is located in the temporo-parietal junction, which is also responsible for speech perception and language comprehension. The planum temporale is typically larger in the left hemisphere (in particular in the case of right-handers). This left-sided asymmetry was also found for the homolog areas Tpt in chimpanzees. In macaques, area Tpt is involved in multisensory processing, but seems to play a major role in auditory processing such as discrimination of sound location (Smiley et al. 2007; Hackett et al. 2007). It is also known that area Tpt is involved in processing species-species vocalizations in chimpanzees and in Old World Monkeys (Taglialatela et al. 2009). Thus, Wernicke's area may be an older area than Broca's area (POp, pars opercularis; PTr, pars triangularis). Broca's area seemed to have evolved along with the increasing complexity of grammatical dependencies and to coordinate speech and language production. It might be that Australopithecus sedipa, which supposed to have evolved into H. ergaster/H. erectus about 2 mya, was able to refer to individual perceivable objects by using instinct calls. While comprehension of objects was supported by multisensory concept formation, the ability to communicate about these experiences presumably relied at this stage on sound imitations replacing instinct calls. The model here implies that mothers were singing to their babies by imitating sounds. Moreover, these sound imitations were successively amended by syllabic consonant-vowel (CV) patterns (e.g., *ma-ma, ah-ah*). These patterns were part of rhythmic melody to comfort and reassure the offspring. To what extent other forms of expressions such as gestures or dance has been used in combination with vocalization for clarifying the expressed meaning, remains an open question. However, protomusic is here the hub of the present account and we assume that syllabic combinations were used to refer to objects (e.g., *he-he ro-ro* >>danger lion). The examples here solely serve as illustrations, but do not indicate the attempt of a realistic approximation of the vocalized patterns. The next milestone protophrases involves the ability to use discrete units by using particular patterns as general iconic attributes for a variety of different objects; *he-he*, for example, could have been combined with any objects that seem to indicate danger (e.g., *he-he na-na* >>danger strangers). Although these protophases may have their roots in protomusical patterns, their units can be considered as discrete units, which can be infinitely combined and represent basis syntactic structures. But we do not make claims that different units were systematically combined at this stage to form hierarchical syntactic structures. This function, called merge, might have been developed in its basic form during the epoch of H. erectus.

H. erectus, assumed to be the direct descendent of Au. sediba, may be the species, which began to use these syllabic combinations. Along with the development towards embodied concepts, a large set of phrases were developed stimulating an increase of working memory capacities in the prefrontal cortex. It has been discussed whether the function merge were born out of syllabic structures. Carstairs-McCarthy (1999) discusses an evolutionary model of syllable frames, which are considered as the foundation for initial syntax. Although this model faced some criticism (e.g., Tallerman 2005), still some aspects of this approach seemed to be plausible for the development of linguistic structures.

(1)

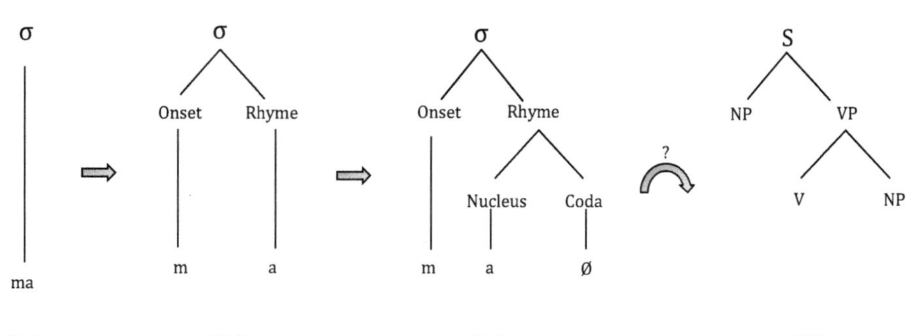

First, during the time span of ca. 1.8 million years, H. erectus developed this vocal communication system to label ideas. An elaborated conceptual system with abstract and symbolic meanings referring to general concepts away from the immediate sensory experience was created. One important aspect of this development refers to the ability to use discrete units not only at the linguistic level. The function merge with the outcome to generate hierarchical structures for the purpose to cluster complex computations may have been developed at the cognitive level. Thus, a plausible scenario might be that merge was not the result of syllabic computations *per se* that were then in turn mimicked for syntactic analysis, as shown in (1), but represented (and represents) an underlying cognitive ability used at different linguistic and non-linguistic levels: increase of syllabic frame complexity (1a.–c.), whereas (1c.) were mimicked by (1d.) to generate syntactic frames. Thus, we assume here that H. erectus used protospeech to communicate. As the phrases presumably did not consist of more than three discrete units, as described in more detail below, the scope of syntactic computations were limited compared to modern language. However, there are no reasons to assume that the number of discrete lexical units was somehow restricted. While the syllabic frame account might be a plausible scenario, it does not make any assumptions about the neurocognitive changes associated with the development of word order processing (linguistic syntax). Broca's area and associated subcortical structures are primarily involved in linguistic syntax as we discuss otherwise in more detail. Most interesting, Broca's area is also involved in processing music syntax at different timescales (Patel 2003). In ancient history of primates, before the development of protospeech and modern language, Broca's area may have its genetic roots in efficient computations of motor sequences. It is not only connected with parietal-temporal-occipital regions but also with the subcortical structure such as the reward circuits to express meanings. In the case of music, studies indicate that the auditory sensory cortices collaborate with mesolimbic regions (in particular with the nucleus accumbens, NAcc) to generate a rewarding stimulus. Highly desirable items produce an enhanced NAcc connectivity with the regions of the inferior frontal gyrus (IFG). Based on a diffusion-weighted imaging (DWI) fiber tracking method, direct in vivo evidence for a structural connectivity between Broca's area and the bansal ganglia/thalamus was recently reported (Ford et al. 2013; Salimpoor et al. 2013). The NAcc, the reward system, is part of the basal ganglia and if a pleasure sensation is generated in context of tasty food, love, win, or pleasant music, just to name a few examples, more dopamine will be released into the ventral tegmental area, NAcc and the prefrontal cortex. Thus, the neural circuits connecting the basal ganglia with the prefrontal cortex, including Broca's area, is an important functional interface for enjoying and computing music in modern humans. Moreover, Broca's aphasic patients, who primarily suffer from syntactic processing deficits, usually do not have only lesions in Broca's area but the lesions involve also subcortical structures. Thus, lesions to Broca's area alone do not cause Broca's aphasia, but the neural circuits associated with Broca's area. While we do not discuss here the shared or common neural resources of music and language, we assert that music and linguistic phrasing recruit the same neural circuits.

Similar to the evolution from H. erectus to H. sapiens, the evolution from protospeech to modern language occurred gradually, at least there is no plausible reason to assume otherwise. Among other significant changes, an increase of working memory span in the prefrontal cortex is one important condition for processing more complex linguistic syntax that involves non-canonical sentence structures or long-distance dependencies. At the same, sub-lexical processing continued to evolve towards lexical morphology to mark different meanings. Today's modern languages using the following morphological structures, which all were presumably not components of protospeech: affixes (e.g., Case, Gender, Number, Time), lexical derivations (e.g., word class changes: verb to noun), irregular verbs (e.g., separate lexical entry instead of using a past tense rule for regular verbs), contrast between function words (typically: determiners, pronouns, conjunctions, pre- and postpositions), and content words (typically: nouns, verbs, adjectives, adverbs). Furthermore, the ability of figurative speech evolved reflected in various different types such as metaphors, fixed phrases (including idioms), puns, humor, irony, neologism, hyperboles, satire, etc.

Simultaneously, protomusic evolved to modern music abilities and it is to assume that other H. sapiens' specific cognitive capacities such as fine motor control of hands and feet evolved to those abilities we use today on a daily basis. The literature about a connection between language processing and the reward system is virtually non-existent, but this thematic niche may require experimental studies. Successful communication as any other cognitive computation is certainly rewarding and is inter alia an emotive act, in particular if the dialog partner responds as desired. Our ancient reward and motivational system may provide the answer on how original harmonic sound patterns gradually transformed to the first linguistic computations.

References

Arbib, M. A. (2005). From monkey-like action recognition to human language: an evolutionary framework for neurolinguistics. *The Behavioral and Brain Sciences, 28*(2), 105–124 and discussion 125–167.

Bickerton, D. (1990). *Language and species*. Chicago: Chicago University Press.

Bickerton, D. (2003). Symbol and structure: A comprehensive framework for language evolution. In M. H. Christiansen & S. Kirby (Eds.), *Language evolution* (pp. 77–93). Oxford: Oxford University Press.

Bickerton, D. (2007). Language evolution: A brief guide for linguists. *Lingua, 117*(3), 510–526.

Brown, S. (2000). Evolutionary models of music: From sexual selection to group selection. In F. Tonneau & N. S. Thompson (Eds.), *Perspectives in ethology. 13: Behavior, evolution and culture* (pp. 231–281). New York: Plenum Publishers.

Call, J., & Tomasello, M. (Eds.). (2007). *The gestural communication of apes and monkeys*. Hillsdale: Lawrence Erlbaum.

Carstairs-McCarthy, A. (1999). *The origins of complex language: An inquiry into the evolutionary origins of sentences, syllables and truth*. Oxford: OUP.

Corballis, M. C. (2003). From mouth to hand: Gesture, speech, and the evolution of right-handedness. *Behavioral and Brain Sciences, 23,* 199–260.

Darwin, C. (1859). *On the origin of species by means of natural selection, or the preservation of favoured races in the struggle for life*. London: John Murray.

Darwin, C. (1871). *The descent of man, and selection in relation to sex*. London: John Murray.

Egnor, S. E. R., & Hauser, M. D. (2004). A paradox in the evolution of primate vocal learning. *Trends in Neurosciences, 27*(11), 649–654.

Feakins, S. J., Levin, N. E., Liddy, H. M., Sieracki, A., Eglinton, T. I., & Bonnefille, R. (2013). Northeast African vegetation change over 12 m.y. *Geology, 41*(3), 295–298.
Fitch, W. T. (2000). The phonetic potential of nonhuman vocal tracts: Comparative cineradiographic observations of vocalizing animals. *Phonetic, 57,* 205–218.
Fitch, W. T. (2013). *Musical protolanguage: Darwin's theory of language evolution revisted. In Birdsong, speech, and language: Exploring the evolution of mind and brain.* Cambridge: MIT Press.
Ford, A. A., Triplett, W., Sudhyadhom, A., Gullett, J., McGregor, K., Fitzgerald, D. B., Mareci, T., White, K., & Crosson, B. (2013). Broca's area and its striatal and thalamic connections: a diffusion-MRI tractography study. *Frontiers in Neuroanatomy, 7*(8). doi:10.3389/fnana.2013.00008.
Hackett, T. A., De La Mothe, L. A., Ulbert, I., Karmos, G., Smiley, J., & Schroeder, C. E. (2007). Multisensory convergence in auditory cortex, II. Thalamocortical connections of the caudal superior temporal plane. *The Journal of Comparative Neurology, 502*(6), 924–952.
Hewes, G. W. (1973). Primate communication and the gestural origin of language. *Current Anthropology, 14,* 5–24.
Holloway, R. L. (1969, 1992). Culture: A human domain [reprint 1992]. *Current Anthropology, 10*(4), 47–64.
James, W. (1890). *The principles of psychology* (Vol. 1). New York: Henry Holt [reprint 1999: Bristol: Thoemmes Press].
Jespersen, O. (1922). *Language: Its nature, development and origin.* New York: W.W. Norton & Co
Miller, G. (2000). Evolution of human music through sexual selection. In N. L. Wallin, B. Merker, & S. Brown (Eds.), *The Origins of Music* (pp. 329–360). Cambridge MA: MIT Press.
Müller, F. M. (1866). *Lectures on the science of language: Delivered at the Royal Institution of Great Britain in April, May, & June 1861.* London: Longmans, Green.
Patel, A. D. (2003). Language, music, syntax and the brain. *Nature Neuroscience, 6*(7), 674–681.
Pinker, S. (1997). *How the mind works.* New York: W.W. Norton & Co Inc.
Salimpoor, V. N., Bosch, I. van den, Kovacevic, N., McIntosh, A. R., Dagher, A., & Zatorre, R. J. (2013). Interactions between the nucleus accumbens and auditory cortices predict music reward value. *Science, 340*(6129), 216–219.
Smiley, J. F., Hackett, T. A., Ulbert, I., Karmas, G., Lakatos, P., Javitt, D. C., & Schroeder, C. E. (2007). Multisensory convergence in auditory cortex, I. Cortical connections of the caudal superior temporal plane in macaque monkeys. *The Journal of Comparative Neurology, 502*(6), 894–923.
Spencer, H. (1857). On the origin and function of music. Essays on education and kindred subjects *Fraser's Magazine*.
Taglialatela, J. P., Russell, J. L., Schaeffer, J. A., & Hopkins, W. D. (2009). Visualizing vocal perception in the chimpanzee brain. *Cerebral Cortex, 19*(5), 1151–1157.
Tallerman, M. (2005). Initial syntax and modern syntax: Did the clause evolve from the syllable? In M. Tallerman (Ed.), *Language origins. Perspectives on evolution* (pp. 133–152). Oxford: OUP.
Tallerman, M. (2007). Did our ancestors speak a holistic protolanguage? *Lingua, 117*(3), 579–604.
Thomas, D. A. (1995). *Music and the origins of language: Theories from the French enlightenment.* Cambridge: Cambridge University Press.
Wray, A. (1998). Protolanguage as a holistic system for social interaction. *Language & Communication, 18*(1), 47–67.

Chapter 3
Genetic Foundations

3.1 Language-Related Genes

Genes carry information to build and maintain cells of an organism and are passed to its offspring. Genes refer to a particular stretch of DNA (deoxyribonucleic acid) and RNA (ribonucleic acid). Typically, the information travels from DNA to RNA and then to proteins (chains of amino acids), which effects the structure and function of an organism. Thereby, a transcription factor often plays an important role in development and intercellular processes. A transcription factor is a protein, which binds to a specific DNA sequence and regulates the transcription process of genetic information from DNA to mRNA (messenger RNA). Thus, the birth of a new life is a coded biochemical package including information about development, regulation, and maintenance of cellular structures that also determines the neural circuits for language processing. Some disorders are associated with mutations of transcription factors. It is important to realize that the absolute number of genes a single organism carries is not responsible for the differences between organisms. For instance, the single-celled organism "yeast" (used in beer or bread) has not got much fewer genes than a multicellular organism such as a human. One single gene does not only produce numerous proteins, but also will be turned off and on by sequences in a specific way for each organism. Although chimpanzees and humans have approximately 99% of their genes in common, a DNA sequence of 1% consists of about three billion codes (letters). It is estimated that since the split between apes and humans (ca. 7–5 mya), 15 million letters were changed which were specific to humans.

In 2004, the human genome project (HGP) was completed. Approximately 23,000 genes (3.2 base pairs long) present in modern humans and how they are expressed provide crucial information about the relationship between genotype and phenotype, including the biological disposition of language (BDL). It can be considered as a combined reference genome of a small number of donors. In 2012, the 1000 Genomes Project was finalized for mapping human genetic variation. Such projects and other related projects will certainly drive our knowledge of the BDL in the near future. For instance, the Neanderthal genome project (NGP) that was large-

D. Hillert, *The Nature of Language,* DOI 10.1007/978-1-4939-0609-3_3, ly completed in 2010, showed that 99.7% of the base pairs are shared by modern humans and Neanderthals. The speech-related gene "FOXP2 transcription factor," which will be further described in detail, showed the same mutations in the Neanderthal species 1,253 and 1,351c as in modern humans ($Foxp2^{neander} = Foxp2^{hum}$, Krause et al. (2007).[1] These results seem to be valid, as different techniques were applied meanwhile to address possible contaminations with human DNA. The analysis of the common chimpanzee genome was published in 2005 by the Chimpanzee Sequencing and Analysis Consortium. Humans, in contrast to Hominidae (great apes), have 23 pairs of chromosomes instead of 24 pairs. Chromosome 2 evolved in humans from two small chromosomes (2A and 2B). Thus, parts of chromosome 2 are scatted in vertebrates that are more distantly related to humans (e.g., rodents). About 2.7% genome difference is the product of gene duplications or deletions. It was found that not only the above mentioned FOXP2 is different in humans, but also the hearing gene alpha tectorin. Chimpanzees' alpha tectorin seems to be less sensitive to nuances of human speech perception (Clark et al. 2003). Besides FOXP2, the genes ASPM, microcephalin and GLUD2 were associated with language and GNPTAB, GNPTAG, and NAGPA with stuttering. At present, it is estimated that GLUD2 appeared 23 mya when hominoids (apes and humans) split from Old World monkeys. It seems that GLUD2 contributes to the development of cognitive abilities and language skills. It is present only in humans and apes. It regulates glutamate and thus the communication between the neurons.

Moreover, it is thought that an abnormal spindle-like microcephaly-associated gene (ASPM) and microcephalin (MCPH) have contributed to the increase in cortical structures. Both genes are present in humans and chimpanzees, and mutations in either result in small brains and mental disorders in humans. A more specific role of the variants ASPM-D and MCPH-D in acquiring a particular linguistic typology has been examined by Dediu and Ladd (2007). First, they correlated 983 genetic variants, including those mentioned above, with 26 typological variants (e.g., the amount of consonants, inflections, and tones) used in languages by 49 different populations. The correlation findings were negative. However, ASPM-D and MCPH-D significantly correlated with speaking a nontonal language.[2] The authors speculate that these genetic variants are in charge of subtle cortical structures associated with tone perception. Populations with a high frequency of these variants would have possibly developed more non-tonal languages. This interpretation remains highly speculative, in considering that correlations do not provide causal relations. The possibility of a genetic bias developed through many generations does not prevent us from acquiring any linguistic features due to our neurocognitive plasticity.

Again stuttering, which is a speech dysfluency disorder, can have multiple reasons. However, mutations associated with the three genes GNPTAB, GNPTAG, and NAGPA seem to be the reason for stuttering in only 9% of the

[1] The FOXP2 gene of the El Sidrón 1253 and 1351c specimen shows the same mutations at positions A-911 and G-977 in exon 7 as in modern humans.

[2] Tonal languages use tones (tonemes) to distinguish words or inflections (e.g., Bantu languages, Chinese, Vietnamese, Thai).

cases (Kang et al. 2010). In particular, these mutations seem to affect the lysosomal enzyme targeting the pathway in persistent developmental stuttering. If there is a genetic link, stuttering runs in families. Kang et al. referred to a large Pakistani family with 46 members, who had alterations of the single gene GNPTAB. Several other genes associated with GNPTAB were also defective. This type of mutation was also reported for 77 unrelated Pakistanis and in a group of American and British stutterers. Moreover, based on these findings it has been emphasized that additional neuroimaging work would further bring light to the cause(s) of this class of stuttering (Büchel and Watkins 2010). Mutations in GNPTAB (alpha and beta subunit) and GNPTAG (gamma subunit) of GlcNAc-1-phophotransferase as well as the evidence of affected lysosomal functions would represent the neurochemical basis for atypical white matter structures, documented by diffusion tensor imaging (DTI)[3]. White matter abnormalities would be therefore related to a disconnection syndrome responsible for stuttering.

3.2 The Role of the Basal Ganglia

In the 1990s, members of a well-known British family "KE," the members of which had suffered for three generations from relatively severe speech and language disorders, were studied (e.g., Hurst et al. 1990; Fisher et al. 1998, 2003). The speech was dysfluent and sentence production was agrammatical. However, language comprehension appeared to be preserved and IQ was at the lower end of the normal range (e.g., Vargha-Khadem et al. 1995). In general, this disorder was considered as a speech output problem, but not as grammatical or broader language disorder (but see Gopnik 1990). An unrelated single case (CS) was then reported, who had the same speech deficits as the members of the KE family. It was found that the mutation was the result of an arginine-to-histidine substitution at position 553, called R553H (Lai et al. 2001). R553H as part of the FOXP2 transcription factor represents a loss-of-function mutation and is conserved across the KE family members. While the primary deficit can be related to the difficulty of acquire complex muscle movements to form and combine sounds in words and sentences, secondary consequences might affect language processing as well as language-related cognitive processes. Magnetic resonance imaging (MRI) studies conducted on the affected individuals of the KE family, revealed a bilateral volume reduction of 25% of the caudate nucleus and the basal ganglia, which are both motor-related neuroanatomical structures. The caudate nucleus significantly correlates with oral apraxia tests.[4]

[3] White matter connections can be better analyzed with DTI and fiber tractography than with standard MRI. The DT-MRI method measures in vivo and non-invasively the random motion (diffusion) of hydrogen atoms within water molecules (or other moieties) in all three dimensions. Water in tissues, which consist of a large number of fibers such as brain white matter, and DT-MRI renders in 3D complex information how water diffuses in tissues.

[4] Damage to the basal ganglia can result in motor disorders such as Parkinson disease, Huntington disease, Tourette syndrome, obsessive-compulsive disorder. The caudate nucleus is located within the basal ganglia.

Interestingly, the caudate nucleus seems to be closely connected to the thalamus, which can be considered as a switchboard for multiple functions. These functions also include the control of articulation when switching between two different languages. A gray matter reduction was found in Broca's area while the temporal gyrus showed increased gray matter volume. In addition, the somatosensory and occipital cortices were affected, although these structures are not directly related to motor, speech, or language functions. However, members of the KE family were able to compensate for morphologically complex words by recalling the whole lexical form or by using particular rules taught to them. Thus, despite its genetic cause, this disorder is treatable.

Numerous comparative studies were conducted to explore the specific role of FOXP2 in other species to draw conclusions about this transcription factor in human speech and language processing. Since Foxp2^{hum} plays such an important role in speech, different kind of comparative studies were conducted. For instance, the behavior of biologically engineered mice with Foxp2^{hum} was different from normal control mice. The Foxp2^{hum} mice produced altered ultrasound squeaks, were less fearless and learned more quickly to use visual and tactile clues to solve a maze. Their brains had more and longer dendrites (afferent biochemical connections among neurons) and increased synaptic plasticity than normal mice (Enard et al. 2009). Interestingly, the effect of quicker learning relied on just one of the two amino-acid changes in the Foxp2^{hum} while the other mutation seem to have no effect. In a different study, two different FOXP2 versions were inserted in vitro in human brain cells to find out more about the expression of the genes. Compared to Foxp2^{chimp}, the human version increased the expression of 61 genes and decreased the expression of 51 genes (Konopka et al. 2009). They conclude that the data indicate different transcriptional regulations and are not due to different levels of both versions. Some of the genes can be linked to motor aspects of speech and to cortical and craniofacial development. Thus, FOXP2 may regulate the development of neural and physical structures involved in speech. Moreover, in fetal macaques, FOXP2 expresses in the basal ganglia (as in the members of the KE family) and in mice in various structures including thalamus, hypothalamus, and cerebellum (Fig. 3.1).

The role of the basal ganglia in language processing is well known. For instance, aphasia always involves damage in the basal ganglia, and in the case of Parkinson disease, the laryngeal control degenerates when the basal ganglia are affected. Thus, much evidence speaks against Fitch's (2006) assumption that direct cortico-laryngeal brainstem circuits (if they exist) are solely critical for the laryngeal activity as language production also involves the coordination of lips, tongue, and breathing.

In general, there seem to be a significant evolutionary diversity in FOXP2 sequences. These are orthologously expressed in brains of amphibians, fish, and birds. For instance, the male zebra finch gradually matches its song skills to the tutor's song, usually its father. FOXP2 seems to be involved in the development of the song system of zebra finches. FOXP2 has been also sequenced in species, which vocalize and echolocate (e.g., bats, elephants, whales, dolphins). However, at present it is difficult to draw direct connections between sequence changes and vocal learning skills. Similarly, the particular role of FOXP2 in human remains vague. There is no

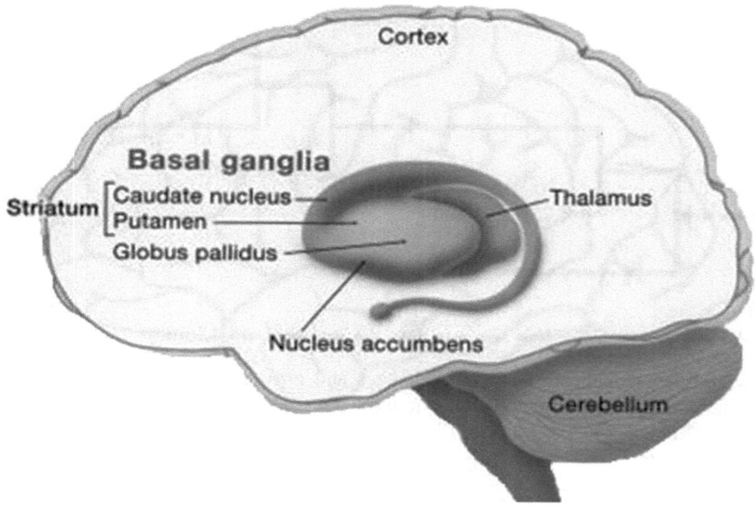

Fig. 3.1 The basal ganglia consist of the caudate nucleus, putamen, and globus pallidus. The cortico-basal ganglia circuits involve motor sequences for walking/running, speaking, and sentence comprehension. (Adapted and modified, Lieberman 2009; @ Elsevier Limited)

doubt that this transcription factor is involved in speech, but the specific amino acid substitutions cannot be determined with respect to speech and/or language. It might be possible that FOXP2 regulates a whole set of genes responsible for the development of specific brain structures, supporting the capacity of speech and therefore language processing.

FOXP2 is not a new human-specific gene. But what enables humans then to use language? Foxp2^{hum} differs from chimpanzee and gorilla sequences at two amino acids (out of 740) and from the orangutan and mouse sequences at three and four residues. Both replacements (T303N and N325S), which probably resulted from positive selection, took place in exon 7: T303N is a threonine-to-asparagine substitution at position 303 and N325S is an asparagine-to-serine substitution at position 325 (Fig. 3.2). They seem to be fixed in humans as they occur in 226 human chromosomes.

Thus, it might be possible that the two replacements evolved in the human lineage are responsible for the BDL. However, FOXP2 is a large transcription factor and besides noncoded regulatory sequences, there are >2,000 differences between Foxp2^{hum} and Foxp2^{chimp}. There is little doubt that, FOXP2 is the only one selected gene in the evolution of the BDL. Mutations of FOXP2, ASPM, or MCPH in humans, for instance, are based on positive selection (Darwinian selection), an evolving process that favors the prevalence of beneficial traits such as speech/language and cognition.

FOXP2 in humans is not a speech- or language-specific gene, but is involved in vocal muscle coordination and other motor functions used by vertebrate species. It is yet to be seen as to what extent Foxp2^{hum} differs from Foxp2$^{non\text{-}hum}$ to specify the differences that contribute to the development of language-related cortical

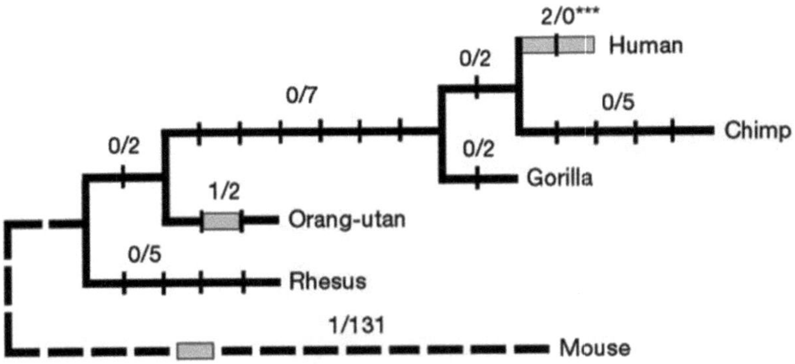

Fig. 3.2 *Tick marks* indicate nucleotide changes and *gray bars* indicate amino-acid changes. The numbers show how many non-synonymous/synonymous changes occurred in each lineage. (adapted, Enard et al. 2002; © Elsevier Limited)

structures. The differences can refer to many factors including the interplay between different genes and/or the specific role of the two amino-acid replacements. Although we just opened the book to start reading the genetic story for language, it might be possible that these FOXP2 differences enable humans to speak, while apes rely on visual communication. In using customized keyboards, the bonobo Kanzi learned up to 500 lexigrams to make requests, answer questions, and compose short phrases, and the number of spoken words he understood was around 3,000. However, the ape's cognitive abilities cannot be compared with those cognitive abilities underlying human language. For example, unlike the human child, bonobos or chimps never ask questions, although they apparently understand the difference between *who, what,* and *where* (Premack and Premack 1983). Apes can learn to associate icons with objects motivated by rewards. Thus, this basic communicative disposition may be part of the human BDL. Research on comparative cognition becomes less human-focused as new discoveries about cognitive skills in animals use new techniques to adapt to the need of a particular species. For instance, Ayumu, a trained juvenile chimpanzee at the Primate Research Institute of Kyoto University, outperformed nontrained adult humans in duplicating the lineup of five numbers (between one and nine) displayed for 210 ms (Inoue and Matsuzawa 2007). Ayumu showed an accuracy level of 79 %. However, when humans were trained on this task, their performance matched or was even better than of Ayumu (Silberberg and Kearns 2009; Cook and Wilson 2010). The findings still indicate that the spatial working memory capacity is well developed in chimpanzees. But, chimpanzees also modify the branches in wildness by breaking off one or both ends, sharpen the stick, and jab the bludgeon into hollows in tree trunks to retrieve bush babies. Bonobos, chimpanzees, orangutans, gorillas, bottlenose dolphins, orcas, elephants, and European Magpies[5] pass the mirror test, which provides a hint for a certain degree

[5] The European Magpie (Pica pica) belongs to the crow family and is believed to be one of the most intelligent animals. Although the European Magpie has no neocortex, the nidopallium (a region of the avian brain responsible for executive and cognitive functions) has a relative size comparable to the neostriatum of the basal ganglia in humans, chimpanzees, and orangutans.

of self-awareness—humans are able to recognize themselves usually at the age of 18 months. New discoveries reveal more and more that animal cognition is much more sophisticated than previously assumed.

However, linking genes with specific traits is still a difficult enterprise. Even the specific role of FOXP2 in speech and language is unclear. While we know that mutations of FOXP2 lead to speech disorders, a direct connection between amino-acid replacements in FOXP2 has not been discovered. FOXP2 replacements in human evolution involve T303N and N325S, while the mutation found among the members of the KE family refers to R553H. FOXP2 might regulate a set of different genes responsible for speech and language or it might regulate cell behavior not only specialized for language processing, but also for other functions. In sum, the broad range of FOXP2 gene expression presumably affects non-speech/language related systems. Thus, the principles of epistasis and pleiotropy also seem to apply in the case of speech and language. The principle of epistasis says that, most of the phenotypic traits are created by means of multiple gene interactions involving modifier genes; the principle of pleiotropy states that most genes affect multiple phenotypic traits simultaneously.

As in the case of any human-specific trait, it is difficult to draw a direct link between genetic changes in evolution and new phenotypes. With respect to language, it also seems to be important to consider that we would be primarily interested in revealing the BDL, that is, the universal characteristics underlying each single language. In considering the complexity of the gene–phenotype link, it may seem more suitable to consider biochemical and physiological characteristics at the cellular level without ignoring the big picture—that is, different levels of investigation from gene regulations to behavioral patterns. Future research directions on the genetics of language may consider, in addition to comparative genomic and microarray studies[6] to apply simulation applications, at best integrating gene expression microarray and neural data. A different kind of basic research approach would imply mathematics to find correspondences between biological, neural, and linguistic principles and rules. The mathematical description at each level alone, has no explanatory power for predicting cognitive behavior in relation to biological processes. Mentalistic or cognitive linguistic[7] attempts to capture universal cognitive principles of syntax and semantics. In an ideal world, our goal would be to find a formal language to integrate linguistic, neural, and biological information. However, a more realistic approach would be an attempt to cross the boundaries of a single level approach. This crossing might be vertical between different levels (e.g., linguistic–neural or molecular–neural) or horizontal between two or more domains on the same level (e.g., syntax of language and music).

---

[6] DNA microarrays measure expression levels of thousands of genes simultaneously. Most microarray systems measure different types of mRNA molecules in cells and thus indirectly measure the expression levels of the genes responsible for the synthesis of those mRNA molecules.

[7] The approach of computational linguistics is, in contrast to mentalistic or cognitive linguistics, exclusively formalistic for the purpose of creating software applications independent of the question "how human cognition works." This does not exclude the possibility that some formalism turns out to be a useful by-product for describing language and cognition in humans.

The genetic viewpoint illustrates how difficult it is to draw a direct link between genotype and phenotype. These attempts are driven by the idea that there has to be an innate structure (sometimes called universal grammar) that allows each child to acquire a language without explicit instructions. At least, we can claim that this innate structure must refer to a symbolic acquisition algorithm (SAA) that enables modern humans to learn thousands of different languages and symbolic systems. At this point, we do not ask questions about the relationships between language and other cognitive domains such as mathematics or music, but we imply that SAA is universal with respect to symbolic processing. Genetic information may trigger the SAA that typically enables humans to acquire languages effortlessly.[8] Thus, SAA must meet a certain search space, that is, it has to be general enough to cover at least all single languages (existing and dead). A major approach in linguistic research is to describe common properties of existing human languages (e.g., Principles and Parameters Theory; Chomsky 1995). According to this model, the child is equipped with innate principles and it sets specific parameters for a particular language. The child intuitively extracts linguistic rules when processing data (linguistic input). Language and symbolism is a complex trait and the positive selection of this trait presumably involves several incremental steps. In the context of Darwinian's paradigm, improved communication skills contribute to fitness and over time, offspring inherit mutations favoring the BDL reflected in the SAA. Thus, a biological model of our language capacity should include the evolution of the genetic foundations for the SAA that enables children to acquire instinctively one or more languages. To what extent the SAA changes and is changing needs to be addressed, but currently we are still in a preliminary stage of relating phylogenetic and ontogenetic properties of language processing to brain structures and behavior.

References

Büchel, C., & Watkins, K. E. (2010). Genetic susceptibility to persistent stuttering. *The New England Journal of Medicine, 362*(23), 2226–2227.

Chomsky, N. (1995). *The minimalist program*. Cambridge: MIT Press.

Clark, A. G., Glanowski, S., Nielsen, R., Thomas, P., Kejariwal, A., Todd, M. J., Tanenbaum, D. M., Civello, D., Lu, F., Murphy, B., Ferriera, S., Wang, G., Zheng, X., White, T. J., Sninsky, J. J., Adams, M. D., & Cargill, M. (2003). Positive selection in the human genome inferred from human-chimp-mouse orthologous gene alignments. *Cold Spring Harbor Symposium on Quantitative Biology, 68,* 471–477.

Cook, P., & Wilson, M. (2010). Do young chimpanzees have extraordinary working memory? *Psychonomic Bulletin & Review, 17*(4), 599–600.

[8] As for any species, there have to be innate restrictions with respect to cognitive behavior. We do not discuss at this point the scope of these restrictions whether biolinguistic approaches, for example, are too restrictive by considering only specific linguistic levels of descriptions such as syntax and/or semantics.

Dediu, D., & Ladd, D. R. (2007). Linguistic tone is related to the population frequency of the adaptive haplogroups of two brain size genes, ASPM and microcephalin. *Proceedings of the National Academy of Science, 104*(26), 10944–10949.

Enard, W., Przeworski, M., Fisher, S. E., Lai, C. S., Wiebe, V., Kitano, T., Monaco, A. P., & Pääbo, S. (2002). Molecular evolution of FOXP2, a gene involved in speech and language. *Nature, 418*(6900), 869–872.

Enard, W., Gehre, S., Hammerschmidt, K., Hölter, S. M., Blass, T., Somel, M., Brückner, M. K., Schreiweis, C., Winter, C., Sohr, R., Becker, L., Wiebe, V., Nickel, B., Giger, T., Müller, U., Groszer, M., Adler, T., Aguilar, A., Bolle, I., Calzada-Wack, J., Dalke, C., Ehrhardt, N., Favor, J., Fuchs, H., Gailus-Durner, V., Hans, W., Hölzlwimmer, G., Javaheri, A., Kalaydjiev, S., Kallnik, M., Kling, E., Kunder, S., Mossbrugger, I., Naton, B., Racz, I., Rathkolb, B., Rozman, J., Schrewe, A., Busch, D. H., Graw, J., Ivandic, B., Klingenspor, M., Klopstock, T., Ollert, M., Quintanilla-Martinez, L., Schulz, H., Wolf, E., Wurst, W., Zimmer, A., Fisher, S. E., Morgenstern, R., Arendt, T., de Angelis, M. H., Fischer, J., Schwarz, J., & Pääbo, S. (2009). A humanized version of Foxp2 affects cortico-basal ganglia circuits in mice. *Cell, 137*(5), 961–971.

Fisher, S. E., Lai, C. S. L., & Monaco, A. P. (2003). Deciphering the genetic basis of speech and language disorders. *Annual Review of Neuroscience, 26*(1), 57–80.

Fisher, S. E., Vargha-Khadem, F., Watkins, K. E., Monaco, A. P., & Pembrey, M. E. (1998). Localisation of a gene implicated in a severe speech and language disorder. *Nature Genetics, 18*(3), 298.

Fitch, W. T. (2006). Production of vocalizations in mammals. In K. Brown (Ed.), *Encyclopedia of language and linguistics* (pp. 115–121). Oxford: Elsevier.

Gopnik, M. (1990). Genetic basis of grammar defect. *Nature, 347*(6288), 26.

Hurst, J. A., Baraitser, M., Auger, E., Graham, F., & Norell, S. (1990). An extended family with a dominantly inherited speech disorder. *Developmental Medicine & Child Neurology, 32*(4), 352–355.

Inoue, S., & Matsuzawa, T. (2007). Working memory of numerals in chimpanzees. *Current Biology, 17*(23), R1004-R1005.

Kang, C., Riazuddin, S., Mundorff, J., Krasnewich, D., Friedman, P., Mullikin, J. C., & Drayna, D. (2010). Mutations in the lysosomal enzyme-targeting pathway and persistent stuttering. *The New England Journal of Medicine, 362*(8), 677–685.

Konopka, G., Bomar, J. M., Winden, K., Coppola, G., Jonsson, Z. O., Gao, F., Peng, S., Preuss, T. M., Wohlschlegel, J. A., & Geschwind, D. H. (2009). Human-specific transcriptional regulation of CNS development genes by FOXP2. *Nature, 462*(7270), 213–217.

Krause, J., Lalueza-Fox, C., Orlando, L., Enard, W., Green, R. E., Burbano, H. A., Hublin, J. J., Hänni, C., Fortea, J., de la Rasilla, M., Bertranpetit, J., Rosas, A., & Pääbo, S. (2007). The derived FOXP2 variant of modern humans was shared with Neandertals. *Current Biology, 17*(21), 1908–1912.

Lai, C. S., Fisher, S. E., Hurst, J. A., Vargha-Khadem, F., & Monaco, A. P. (2001). A forkhead-domain gene is mutated in a severe speech and language disorder. *Nature, 413,* 519–523.

Lieberman, P. (2009). FOXP2 and human cognition. *Cell, 137*(5), 800–802.

Premack, D., & Premack, A. J. (1983). *The mind of an ape*. New York: Norton.

Silberberg, A., & Kearns, D. (2009). Memory for the order of briefly presented numerals in humans as a function of practice. *Animal Cognition, 12,* 405–407.

Vargha-Khadem, F., Watkins, K., Alcock, K., Fletcher, P., & Passingham, R. (1995). Praxic and nonverbal cognitive deficits in a large family with a genetically transmitted speech and language disorder. *Proceedings of the National Academy of Science, 92*(3), 930–933.

Chapter 4
The Rise of Cognition

4.1 Comparative Studies

Modern humans' biological disposition of language (BDL) gradually evolved from nonhuman primates' neurobiological dispositions, whereas some neural structures may have even roots in common ancestry beyond the primate lineage (Darwin 1859). This implies that comparative brain studies reveal homologous language-related areas of the human brain and that those areas can be traced back to cortical structures found in our biological ancestors. It is, thus, not surprising that interdisciplinary research increased toward understanding the evolutionary path of the BDL in modern humans. One line of research uses neuroimaging methods to compare and analyze cortical areas and circuits in different species to draw conclusions about the evolution of these neural structures and about language and cognition (Rilling et al. 2008). A second approach refers to single-cell recordings of the so-called mirror neurons in macaque monkeys (Macaca mulatta) to model the evolution of spoken language in humans (Rizolatti et al. 1996). A third research line focuses on the neural substrate of vocalization primarily in nonhuman primates and in the Avian class to better understand how spoken language works in humans (Bolhuis et al. 2010). In the present chapter, we will provide crucial findings of all three research directions. Finally, a coherent picture will be drawn with the goal to describe which information may be useful for characterizing the cortical properties of the human language system.

In humans, Broca's area[1], the classical motor language area, comprises in the frontal cortex Brodmann areas (BAs)[2] 44 and 45 of the left inferior frontal gyrus (IFG). BA 45 seems to receive more afferent connections from the prefrontal cortex,

[1] Paul P. Broca (1824–1880), a French surgeon and anthropologist, presented in 1861 at the Society of Anthropology of Paris the patient "Leborgne," who was only able to produce the automatism "tan." The autopsy revealed a lesion in the third convolution of the left frontal lobe. According to today's diagnostic methods, he would have been classified as a global aphasic patient. Often, this discovery is considered as the birth of cognitive neuropsychology, although similar observations were made generations earlier by the French neurologist Marc Dax (1836).

[2] In 1909, Korbinian Brodmann (1868–1918), a neurologist from Germany, divided the cortex into 52 distinct cortical regions by considering cytoarchitectonic features.

D. Hillert, *The Nature of Language*, DOI 10.1007/978-1-4939-0609-3_4,

the superior temporal gyrus (STG), and the superior temporal sulcus (STS) and BA 44 from the motor, somatosensory and inferior parietal areas. Thus, different connectivity and cytoarchitecture point to the hypothesis that BAs 44 and 45 perform different functions. The classical receptive language area, called Wernicke's area[3], is also left-sided posterior on the planum temporale (PT), BA 22 (STG; temporoparietal, Tpt), BA 40 (supramarginal gyrus), and sometimes BA 37 is included. The importance of both language centers has been already discovered in the nineteenth century as aphasic patients showed systematic language disorders when the lesion was in one of these centers or in both centers or when the lesion affected the arcuate fascilicus (AF), a long association fiber connecting both centers.[4] Wernicke (1874) already predicted that a lesion to the AF would lead to a different language disorder, known today as conduction aphasia. Aphasic patients with a disconnection syndrome of the AF have particular difficulties with sound processing at the phonemic or syllabic level.[5]

In chimpanzee, the area fronto-orbital sulcus (fo) corresponds to the human BA 44 and may include parts of BA 45 (Jackson et al. 1969; Sherwood et al. 2003). However, the human Broca's cap at the level of the temporal pole is not homologous to chimpanzee's orbital cap. The human Broca's cap includes BAs 45 and 47 rather than BA 44. With respect to the macaque brain, it has been suggested that the inferior limb of the arcuate sulcus includes homologs of BAs 44 and 45 (Deacon 1992; see Fig. 4.1).

A more recent endocast study of the H. erectus Sambungmacan 3 found in Indonesia reported a leftward asymmetry of Broca's cap (Broadfield et al. 2001). This has been taken as evidence for language use in this species. However, there is tremendous variability in humans with respect to the asymmetry of Broca's cap and even leftward asymmetry has been found for BA 44 in three great apes: Pan troglodytes (common chimpanzee), Pan paniscus (Bonobo) and Gorilla gorilla (Cantalupo and Hopkins 2001). Thus, although Broca's area is one of most prominent cortical area involved in language and speech, its leftward asymmetry is not specific to the human cortex and does not inform, therefore, about specific cortical functions associated with this area. Presumably, BA 44 served to coordinate gestures and/or vocalizations in great apes and might have led to an expansion of BA

[3] Carl Wernicke (1848–1905), a German physician resided in Breslau (Wrocław), discovered an aphasic syndrome caused by lesions in the superior temporal lobe with parietal portions. He predicted a third aphasic syndrome based on his and of Paul Broca's discovery: conduction aphasia (original term: "Leitungsaphasie"). Moreover, Lichtheim (1884) developed the so-called Wernicke–Lichtheim diagram to predict four more aphasic syndromes.

[4] Fibers consist of axon bundles that originate from neurons in the (sub)cortical gray matter. Three types of fibers are defined: (1) commissural fibers, which connect the hemispheres; (2) projection fibers, which either connect the cortex to the internal capsule, basal ganglia, brainstem, and spinal cord (corticofugal fibers) or connect the thalamus to the cortex (corticopedal fibers); and (iii) association fibers, which connect adjacent and nonadjacent cortical regions within the same hemisphere and are referred to as short and long association fibers, respectively.

[5] A phoneme is the smallest discrete segmental speech sound (or group of sounds) to form meaningful contrasts between utterances. They can carry stress and tones and can be further decomposed into single phonemic features.

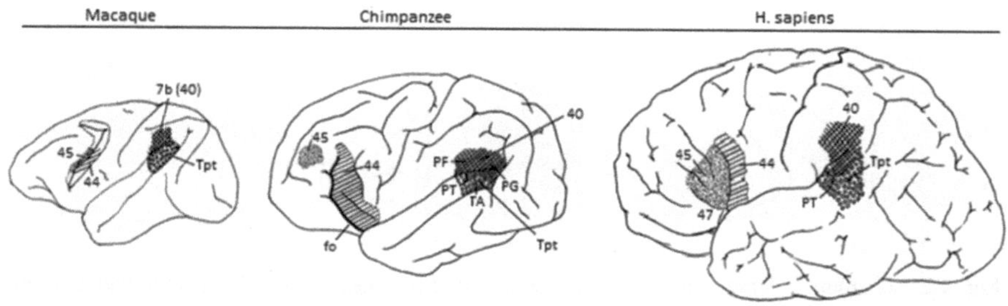

Fig. 4.1 Main left-sided language areas in modern humans and homologs in common chimpanzees and macaque monkeys: BAs 44 and 45 comprise Broca's area; Temporo-parietal (Tpt), planum temporale (PT), and BA 40 comprise Wernicke's area. Human BA 40, chimpanzee PF and PG (inferior and middle parietal lobule), and macaque area 7b are considered as homologs as well as human and macaque areas Tpt and chimpanzee area TA; fronto-orbital (fo) sulcus in chimpanzees. (See text for details; adapted and modified, Falk 2007; © Springer Science+Business Media)

45 and to more cortical folding in the left IFG. Let us turn to the second classical language area. The chimpanzee's homolog of Wernicke's area is the PT. It is larger in the left hemisphere not only in humans but also in chimpanzees (Gannon et al. 1998; Hopkins et al. 1998). Compared to Broca's area, Wernicke's area has been considered as an older cortical structure. One reason is that the Tpt region can be also found in some prosimians (Preuss and Goldman-Rakic 1991).[6] Much like in humans, the left hemisphere is dominant for processing meaningful vocalization in macaque monkeys. Thereby, the STG and the left inferior parietal lobe are homologs of Wernicke's area (Galaburda and Pandya 1982).

Today, different methods are used to map the anatomical connectivity between cortical areas. One reason is that postmortem blunt fiber dissection has its limits when specific fiber systems need to be separated. However, tractographic methods such as diffusion sensor imaging (DTI) is typically used to illuminate fiber pathways in human subjects or electrostimulation in patients prepared for brain surgery. Moreover, in animals histochemical tract-tracing methods are applied, although it is questioned to what extent the human language path can be traced by studying the fiber tracks in nonhuman primates. In addition, it should be emphasized that gray matter is responsible for cognitive functions but not white matter. White matter connects and subserves these cortical areas and has itself not a direct cognitive function. Thus, all these (and other) methods have their limitations. It is, therefore, not surprising that the fiber system for language is controversially discussed (Dick and Tremblay 2012).

However, the core language model considers in analogy to visual processing a dual-stream system: The dorsal route maps auditory input to motoric speech representations and the ventral route maps the auditory input to semantic representations (Ungerleider and Haxby 1994; Hickok and Poeppel 2004; Rauschecker 2011).

[6] Prosimians are primates including lemurs, lorises, bushbabies, and tarsiers and in particular native to Madagascar. Simians are monkeys, apes and humans.

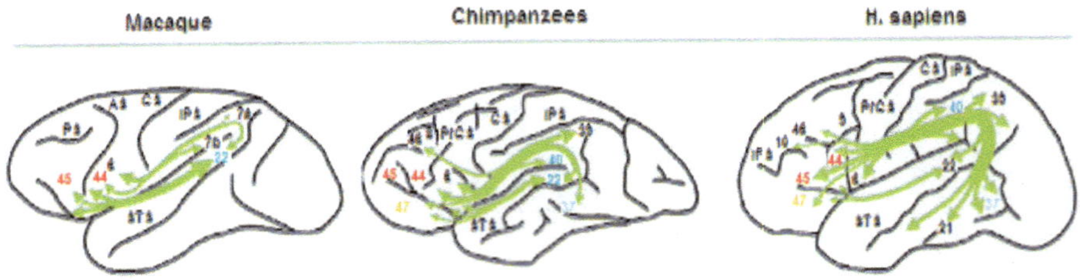

Fig. 4.2 Schematic average tractographic results found for macaque monkeys, chimpanzees, and modern humans. IFS inferior frontal sulcus, IPS intraparietal sulcus, PrCS precentral sulcus, CS central sulcus, STS superior temporal sulcus, PS principal sulcus, AS arcuate sulcus. BAs in red/orange: Broca's area BAs 44 and 45 with extension BA 47; BAs in blue/turquoise: Wernicke's area BAs 22 and 40 with extension BA 37. (Adapted and modified, Rilling et al. 2008; © Nature Publishing Group)

In the case of the dorsal stream, it has been discussed whether there is an additional temporo-parietal–frontal segment (Catani et al. 2005). Glasser and Rilling (2008) suggested dividing the arcuate fasciculus–superior longitudinal fasciculus (AF-SLF) stream between a lexical semantic route from the middle temporal gyrus (MTG) to BAs 44, 45, and 9, and a phonological route from the STG to BAs 44 and 6. In addition, in the right hemisphere, a small stream was found between the superior temporal line (STL) and BAs 44 and 6 and a track connecting the MTG with BAs 44 and 6. The function of the latter route has been attributed to prosodic processing. Moreover, recent studies with nonhuman primates seem to suggest four different segments of the AF–SLF stream, three types of SLF and the AF (Schmahmann et al. 2007; Petrides and Pandya 2006). In particular, the SLF III has been considered as a possible language stream as it connects the inferior parietal lobe with homologs of the human Broca's area (BAs 44 and 45). However, several studies using different methods verify that in macaque monkeys, the area Tpt, which is party considered homologous to Wernicke's area in humans, projects to the dorsal and lateral premotor areas rather than to the language homologs 9/46d, 6d and 8Ad (Petrides and Pandya 2006). Again, significant differences of the neural pathways between Homo sapiens, chimpanzees, and macaque monkeys revealed a comparative DTI study (Rilling et al. 2008; see Fig. 4.2). In humans, the AF strongly connects in the left hemisphere the frontal lobe with the MTG and the inferior temporal gyrus (ITG), including Wernicke's area. In macaques, this region corresponds to the extrastriate visual cortex, next to the primary visual regions. The MTG and ITG enlarged disproportionately in the human lineage following the split of human and chimpanzee lineages. The pathways in chimpanzees are slightly more developed as compared to those in macaques and may reflect some prior conditions that enable the connection between meanings and motor sequences.

However, it is apparent that along with the increase of the human brain size, the white matter volume increased in frontal and temporal areas, in particular the dorsal route. During human evolution, the temporo-frontal circuit mainly expanded for tool and language use. This expansion may have occurred gradually and we can only speculate that the temporo-frontal lobe projections might have been less developed in H. erectus than in H. sapiens but probably more than in Lucy, for instance, who belongs to the genus Australopithecus. However, we cannot exclude the possibility that the fiber stream systems were equally developed in *H. erectus* as in modern humans. The evolution of the AF and SLF is another indication for the gradual development toward the human language capacity and seems not to be the result of a sudden mutation.

Another human diffusion fiber tractography study reveals that the AF runs from the posterior STG to BA 44, sometimes to BA 45, and projects also as in monkeys to the prefrontal BAs 6 and 8 (Frey et al. 2008). This study reported also that the SLF connects in most cases the supramarginal gyrus (SMG) to BA 44. Again, the middle longitudinal fasciculus (MLF) and inferior longitudinal fasciculus (ILF) connect the STL with the ventral posterior intraparietal region, which in turn closes a circuit from the posterior auditory cortex to BA 44. Moreover, as in monkeys, the bidirectional extreme capsule fiber system (ECFS) and the uncinate fasciculus (UF) connects the anterior temporal lobe with BAs 45 and 47. These anatomical models compete with functional models postulating functional routes of specific linguistic processes. Accordingly, it has been proposed that the ventral pathway is used, among others, for mapping sound to meaning, speech recognition, or basic grammar. Magnetic resonance imaging (MRI) data obtained by Buchsbaum et al. (2005) indicate that basic phonological perception seems to be performed by the ventral stream and rehearsal operation are sustained by the dorsal stream.

A similar view has been presented by Hickok and Poeppel (2007), but they discuss in addition a right-hemispheric ventral stream and suggest that the right ventral stream integrates information over longer timescales while shorter timescales might be bilaterally represented. The most critical fiber streams discussed so far are illustrated in Fig. 4.3. One part of the SLF connects a monkey's inferior parietal lobule PF (homolog of anterior superior marginal gyrus, aSMG) with the ventral premotor area BA 6 (green). Another part of the SLF links a monkey's middle inferior parietal lobule areas PFG and PG (homologs of the posterior SMG and angular gyrus, AG) to Broca's area (BAs 44 and 45) and the AF connects Wernicke's (STS) with Broca's area (both in red). However, it is difficult separating it from the inferior branch of the SLF. The MLF connects the posterior temporal areas with the inferior parietal regions PFG and PG (blue). Finally, a ventral route running via ECFS connects the middle and anterior temporal lobes to BAs 44 and 45 (see also Petrides and Pandya 2009).

Furthermore, using resting state functional connectivity (RSFC) analysis, Kelly and colleagues found a robust dissociation between BA 6 and Broca's area (BAs 44 and 45). BA 6 is mainly involved in orofacial control and seems to be most strongly connected to the inferior part of the parietal lobus and has been confirmed by RSFC and monkey models. In contrast, Broca's area connects to the posterior part of the parietal lobe. This distinction in connectivity between BA 6 and Broca's area has also cytoarchitectonic reasons. BA 6 is agranular (in monkeys and humans), that is, it lacks layer IV of the cortical structure. BA 44 can be considered as dysgranular as layer IV is only rudimentarily developed. BA 45, however, has a well-developed layer IV. Layer IV sends efferent fibers to the thalamus and establishes reciprocal excitatory and inhibitory connections to the thalamus and adjacent areas (Lam and

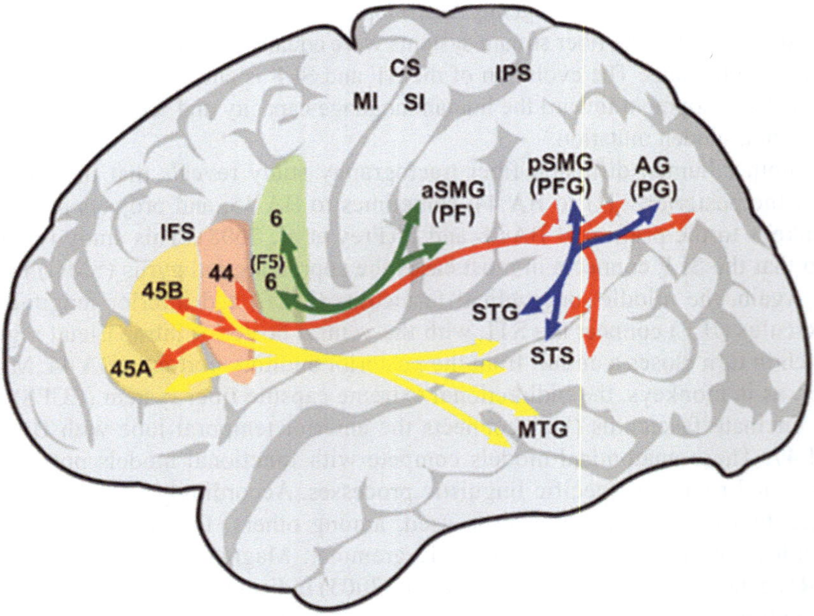

Fig. 4.3 Fiber associations for possible language circuits based on resting state functional connectivity in humans and autoradiographic data in monkeys. CS central sulcus, IPS intraparietal sulcus, MI primary motor area, SI primary somatosensory area, M/STG middle/superior temporal gyrus. (Adapted and modified, Kelly et al. 2010; © Federation of European Neuroscience Societies and Blackwell Publishing Ltd; figure based on © Aboitiz, 2012)

Sherman 2010).[7] RSFC revealed that BAs 44 and 45 are more similar to one another than to ventral BA 6, which has been also confirmed by connectivity studies of the homologs in monkeys (Petrides and Pandya 2009). However, BAs 45 and 44 seem to have connectivity differences. Monkey studies show that PG and the ventrally adjacent temporal cortex are connected stronger to BA 45 than to BA 44. Similarly, RSFC analysis indicates greater BA 45 connectivity to AG and to the S/MTG relative to BA 44.

However, there are still many open questions concerning the specific tasks and properties of these different fiber streams. Typically, AF is said to be involved in articulatory, phonological, and syntactic processes and the ECFS in semantic processing (Friederici and Gierhan 2013; Catani et al. 2005; Saur et al. 2008). In contrast, it has been also argued that the extent to which fiber pathways are engaged depends on cognitive demand rather than on what kind of linguistic information needs to be processed (Rolheiser et al. 2011).

Moreover, a unique set of neurons have been first discovered in macaque monkeys (Rizzolatti et al. 1996; Gallese et al. 1996; Rizzolatti and Craighero 2004). They recorded single visuomotor neurons in the premotor cortex (area F5) of the

[7] K. Brodmann (1909) divides the neurons of the cerebral cortex into six main layers, from the pia (mater) to the white matter.

macaque. The homolog of a monkey's area F5 is ventral BA 6, which extends to BA 44 and PF in humans (cf. Fig. 4.3). These premotor neurons fired when the macaque performed an action (e.g., reaching for food). Some of the neurons also fired when the macaque observed a similar action by another monkey or by a human (e.g., picking up food). These neurons are called mirror neurons as they only discharge when an action is performed or seen but not when an object is presented without an action. One-third of the F5 mirror neurons are classified as strictly congruent, that is, they discharge when the actions are observed and perform corresponding actions in terms of the goals. Two-third of the F5 mirror neurons are defined as broadly congruent as they discharge during observation without motoric encoding of the same action. Moreover, when a macaque heard the sound of action typically associated with an object, audiovisual mirror neurons fired in F5, but for most neurons the discharge was smaller for sound alone than for vision and sound (Kohler et al. 2002). Also, F5 seems not exclusively related to manual tasks (Jürgens 2003). For instance, in rhesus monkeys, the cortical larynx representations overlap with F5, and BAs 12 and 45 are involved in vocalization (Matelli et al. 1985). In sum, these single-cell recordings in primates revealed that some properties of mirror neurons are located in the premotor and parietal cortices.

More recently, the role of mirror neurons in the evolution of language has been more broadly discussed (Corballis 2003; Arbib 2005). Arbib introduces the mirror system hypothesis, which makes claims about how action mirror neurons might have contributed to the evolution of spoken language. Accordingly, he divides it into seven interdependent evolutionary stages: (1) grasping, (2) mirror system for grasping shared by common ancestors of humans and monkeys, and (3) imitation system for grasping shared by common ancestors of humans and chimpanzees; the next three stages distinguish the hominid line from the great apes: (4) a complex imitation system for grasping, (5) protosign, a manual communication system with an open lexicon, (6) protospeech, which used the underlying mechanism of the protosign system, and (7) coevolution of cognitive and linguistic complexities by changing action–object frames to verb argument structures, to syntax and semantics. Here, we are most interested in the last four stages, which are more directly related to the evolution of language readiness. However, still this scenario is quite speculative. In particular, we assume that the evolution of complex vocalization does not need to depend strictly on gestures as claimed by Corballis or Arbib. Also, the coevolution of cognition and language started at an earlier stage and it is cognition, as we discussed in the previous chapter, which has probably been the driving force for the invention of modality-independent complex hierarchically organized structures.

Thus, the question is where do these complex computations come from? One assumption is that language and cognition co-evolved in both directions, that is, cognitive and linguistic complexities have been regenerated by furthering or improving the expressivity of spoken language. Interestingly, a functional MRI[8] study

[8] The most popular neuroimaging technique among cognitive neuroscientists is MRI. The invention of MRI did not arrive in one step and was the results of a series of accomplishments in physics.

shows that listening to speech activates a superior portion of the ventral premotor cortex, which overlaps with the speech production area (Wilson et al. 2004). We believe therefore that the vocalization system has been co-evolved with the gestural system and there is no apparent reason to claim that structures developed by the gestural system have been step by step transferred to the vocal system (Izumi and Kojima 2004). Most communication in monkeys and apes is vocal (besides body language), it is not necessary to postulate a gestural stage preceding protospeech. We can assume that they are auditory mirror neurons, which play a role in vocal imitation. Similar to a motor model for speech, vocal mirror neurons simulate the auditory input and can be therefore used to compare auditory input with own articulation (Liberman and Mattingly 1985). The basic vocal mirror neuron system may have become more complex by an increase of voluntary control of the vocalization mechanism. This has been probably accomplished by an increase of auditory working memory (aWM) functions in prefrontal areas (Bosman et al. 2004).

The fundamental issue how modern language evolved in humans becomes clearer with concerted efforts to consider comparative data from nonhuman species. Based on comparative studies with avian, mammalian, and reptilian species, different alternative hypotheses were discussed about the evolution of brain pathways for vocal learning (VL; Jarvis and Kass 2006). It is suggested that the auditory pathway of vocal learners such as songbirds and humans was inherited from amniotes, which lived about 320 mya.[9] VL is a process that refers to the ability to acquire vocalization by imitation rather than by instinct. Thus, a species can modify their vocalizations as a result of experience. It is apparent that VL depends on auditory

A description of the methods and mechanisms behind MRI goes beyond the scope of the present chapter and the reader will be referred to adequate tutorials (e.g., Pooley, 2005). But let us briefly summarize some important facts about these important but still developing noninvasive neuroimaging techniques. The most common kind of MRI is known as blood oxygenation level-dependent (BOLD) imaging and credited to Ogawa et al. (1990). Neurons receive energy in form of oxygen by means of hemoglobin in capillary red blood cells. An increase of neuronal activity results in an increased demand for oxygen, which in turn generates an increase in blood flow. Hemoglobin is unaffected by the magnetic field (diamagnetic) when oxygenated but strongly affected (paramagnetic) when deoxygenated. The magnetic field is generated by an MRI scanner, which houses a strong electromagnet. For research purposes, the strength of the magnetic field is typically 3 T (1 T=10,000 G) and is 50,000 times greater than the Earth's field. It is predicted that the spatial resolution at the cell level requires high-field magnets (far > 10 T; Wada et al., 2010). This difference in magnetic properties causes small differences in the MR signal of blood depending on the degree of oxygenation. The level of neural activity varies with the level of blood oxygenation. This hemodynamic response (HDR) is not linear. The onset of the stimulus-induced HDR is usually delayed by ca. 2 s because of the time it takes that the blood travels from arteries to capillaries and draining veins. There is typically a short period of decrease in blood oxygenation immediately after neural activity increases. Then, the blood flow increases not only to meet the oxygen demand, but to overcompensate the increased demand. The blood flow peaks at around 6–12 s, before returning to baseline. In contrast to a relatively good spatial resolution of < 1 mm, the temporal resolution has its limits.

[9] Amniotes are animals that are adapted to survive in a terrestrial environment. They develop a layered environment, in which their offspring can grow regardless of whether they give birth or lay eggs. Amniotes evolved during the Carboniferous period ca. 320 mya from amphibian reptiliomorphs and include today all reptiles, birds, and mammals.

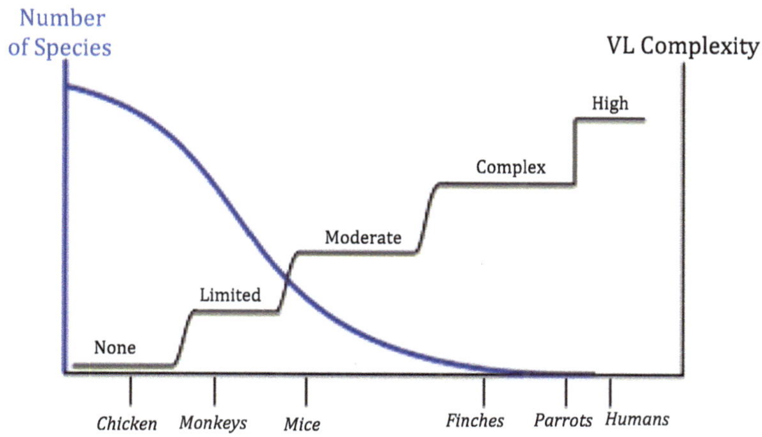

Fig. 4.4 Vocal learning (VL) complexity phenotype. (Adapted and modified; © Petkov and Jarvis 2012)

learning but auditory learning does not depend on VL. For instance, a dog is able to understand a spoken word but cannot reproduce it. Again, vocal learners must be able to use auditory percepts to correct their vocalization. Typical vocal learners closely related to humans are bats, mice, elephants, seals, and cetaceans, and more distantly related are corvid songbirds, some parrots, and mockingbirds. The mammalian as well as the avian group have members which can be considered as vocal nonlearners. However, more recent studies provide evidence that chimpanzees and other primates are also vocal learners, but significantly less capable than, for example, songbirds (e.g., Crockford et al. 2004; Riede et al. 2004; Levréro and Mathevon 2013). Thus, the assumption seems to be obvious that the ability of VL did not evolve from a common ancestor of both groups. But the question is whether particular neural characteristics are shared by both groups that enable vocal modulation. Here, we primarily focus on the anatomical and functional characteristics in birds as they are relative prolific vocal learners compared to other nonhuman vocal learners. The acquisition of human speech and of vocalization in songbirds appears to be comparable as both have sensitive periods before adulthood to acquire and maintain phonological or vocal sequences. Humans are highly prolific vocalizers because they acquire virtually infinite new phonological sequences during their life span. These sequences typically correspond to particular meanings. In contrast, VL in birds such as the male zebra finch is hardly possible after puberty. Their songs have an emotionally rewarding experience such as mating or defending a territory.

Different degrees of VL abilities across species are illustrated in Fig. 4.4; cetaceans are not considered. These data should be considered as an approximation as new studies will certainly reveal that there is more vocal flexibility in some species as previously stated.

VL in humans relies on pathways within the forebrain (prosencephalon). It consists of thalamic structures and the cerebrum (telencephalon), which includes the

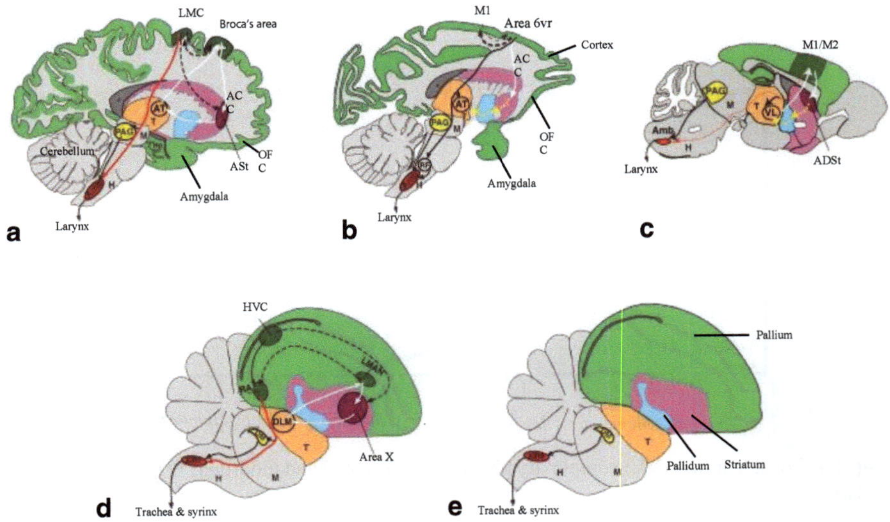

Fig. 4.5 Schematic illustration of the vocal subsystems in humans (a), macaques (b), mice (c), songbirds (d), and chickens (e). ADSt anterior dorsal striatum, Amb nucleus ambiguous, Area 6V ventral part of Area 6 premotor cortex, Area X a song nucleus of the striatum, ASt anterior striatum, AT anterior thalamus, DLM dorsalateral nucleus of the mesencephalon, DM dorsal medial nucleus of the midbrain, H hindbrain, Hp hippocampus, HVC letter-based name, LMAN lateral magnocellular nucleus of the anterior nidopallium, LMC (BA 4) laryngeal motor cortex, M midbrain, M1 primary motor cortex, M2 secondary motor cortex, nXIIts 12th tracheosyringeal motor neurons, PAG periaqueductal gray, RA robust nucleus of the arcopallium, RF reticular formation, T thalamus, VL ventral lateral nucleus of the thalamus. (Adapted and modified, @ Arriaga et al. 2012; see also Petkov and Jarvis 2012)

cerebral cortex and the basal ganglia. In contrast to vocal nonlearning birds, complex VL birds such as songbirds, hummingbirds, and parrots have particular forebrain neuron clusters (nuclei).

It is said that primates have an innate, involuntary vocal production system, which consists of connections from the amygdala, orbitofrontal cortex (OFC) and anterior cingulate cortex (ACC) to the periaqueductal gray (PAG) in the midbrain. Again, PAG neurons synapse to neurons of reticular formation (RF); in turn, they synapse to the α motoneurons in the nucleus ambiguous (Amb), which control the muscles of the larynx for vocal production. In nonhuman primates, BA 6 (Area 6vr) projects in addition to the RF and from there to the Amb. Area 6vr and the ACC are also connected with the primary motor cortex, amygdala, and thalamic structures (not shown). Lesion studies indicate that area 6vr does not control vocalization but respiration (e.g., Jürgens 2009). However, its precise role is unknown, but some preliminary evidence indicates that area 6vr and neighbored areas in the premotor cortex can be modulated during innate vocalization.

Humans rely more on the learned vocal pathway during speech. The learned vocal pathway projects from the face area of the primary motor cortex in BA 4 (laryngeal motor cortex, LMC) to the Amb (Fig. 4.5, red arrow) and a cortico-striatal–thalamic loop for learning vocalizations (two white arrows). This matches in songbirds the direct forebrain projection to vocal motor neurons in the brainstem (robust nucleus of the arcopallium, RA, to XIIts). It has been stated that there is no direct cortico-bulbar projection in vocal nonlearners such as chickens and monkeys. In the case of a lesion in the LMC, learned vocalization is impaired in humans. Thus, it has been assumed that the humans' ability of speech and spoken language is related to the formation of a direct pathway from the LMC to the Amb. In general, studies with nonhuman primates were not able to report evidence for a direct pathway between analogous areas of the LMC (e.g., 6vr) and Amb. A recent study with lab mice, however, might have revealed a region, which seems to be homologous: it is active during vocalization and makes a direct but sparse projection to the Amb (Arriaga et al. 2012; Arriaga and Jarvis 2013). Connections between the anterior forebrain and the posterior vocal motor circuits are marked in Fig. 4.5 with dashed lines. Similar to humans, in mice two pathways converge on Amb, one from PAG and one from the primary motor cortex (M1).

According to new evidence that the vocalization system seems to be more flexible in modulating or mimicking sound patterns in species which were assumed to be nonvocal learners such as mice and monkeys, a continuum hypothesis of VL has been proposed by Jarvis and colleagues. In particular, they may be equipped with an auditory feedback loop. In general, auditory processing can be easier modeled in nonhuman animals compared to learned vocal communication. However, not all patterns of human speech can be modeled in animals and not all vocalization patterns in animals can be modeled in humans. Since a certain group of songbirds and humans are excellent vocal learners, the question arises to what extent the speech and birdsongs can be compared at the syntactic level. With respect to process units, the following match has been proposed between VL songbird vocalization and human speech:

(2)

VL songbirds	Humans
• Song	Song / Speech / Language
• Bout	Sentence
• Motif	Word
• Note / syllable	Phoneme

Figure 4.6 shows an audio spectrogram of a typical zebra finch. Usually, a song starts with some introductory notes followed by one or more motifs. Motifs are repeated sequences of syllables. The question is to what extent songs of VL birds can be compared with spoken language. First, human language consists of a lexicon and syntactic rules. This lexical–syntactic system externally interfaces phonological and phonetic information in perception and production, and internally it interfaces semantics including concepts, intentions, and emotions. Typically, birdsongs do not

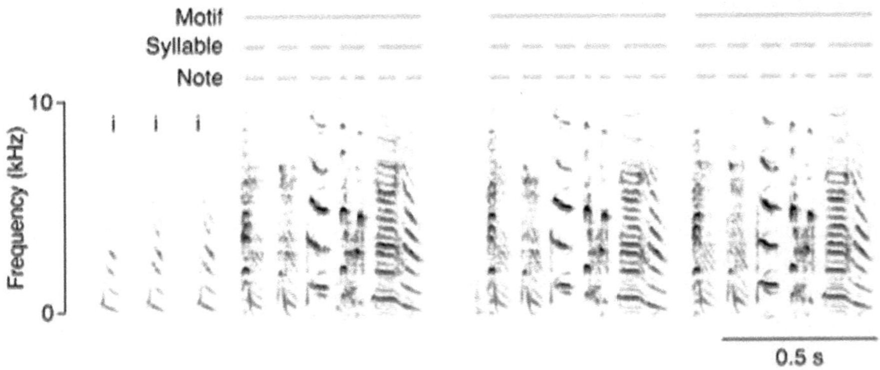

Fig. 4.6 Sonogram of a typical zebra finch song. The introductory notes (i) will be followed by one or more repeated syllables (motifs). Here, a syllable is considered as a continuous sound with one or more coherent time-frequency traces (notes). The continuous rendition of several motifs is called "bout." (Adapted, Berwick et al. 2011; © Elsevier Limited)

express abstract meanings, if at all some intentional or emotive states such as the attempt to attract mates or to defend territories. Thus, we may say that birds (or typically animals in general) use pragmatic meanings rather than semantic meanings. Again, birds make use of a phonetic syntax rather than of a phonological syntax as the term phonology refers to the mental representation of an abstract sound system to convey meanings. In contrast to birdsongs and to animal communication systems in general, human language uses distinct semantic units and morpho-syntactic rules to combine and create open-ended variations of new meanings (see Berwick et al. 2011). In contrast, the communication systems of insects such as bees and ants are innate or hardwired. For example, in a population of honeybees (genus Apis), the forager uses a basic innate compositional signal system to inform nest mates about the location of food and other resources. The orientation of the waggle dance and its duration correlates with the direction and distance that the forager has flown to the food. Thereby, the angle of the waggle dance relative to the upward direction of the comb correlates with the flight direction relative to the sun. Not only dancing seems to play a role but also semiochemicals, sounds, touch, and (comb) vibrations. Sensory data are transformed by innate structures, a process that involves genetic adaption but not learning. Evolution has provided species with a wide range of different communication systems. The overall purpose is to improve survival chances or in other terms to provide pleasure, if we consider food intake as a pleasuring act. Let us return to the learned vocalization abilities of some songbirds.

Birdsongs use variations at the phonetic level but obviously not at the semantic level. For instance, a nightingale can rearrange clusters that generate hundreds of different songs. They may be used for identifying individual birds and express the level of sexual arousal. Canaries use sexy syllables to increase mate attractions, but this does not change the meaning of the song and indicates only the motivation level as Berwick and colleagues point out. On top, humans are able to synchronize the speech-meaning dimension with physical movements (dancing and acting) and/

or complex music patterns (beat and harmonics) for a high-coordinated meaningful and/or pleasurable performance. Since a behavioral comparison is not possible at the semantic level, to what extent is the song syntax of songbirds comparable with the syntax of human language?

Human language consists of nonlinear relationships between words. For instance, in the sentence *The guitarist, who plays bossa nova, went on stage*, the noun phrase (NP) *the guitarist* has to be linked to the verb phrase (VP) *went on stage* to understand the sentence meaning, that is, it is the guitarist, who went on stage. These kinds of nonlinear links are called nonadjacent dependencies, which are inherently part of the hierarchically organized structure of syntax in human languages. Moreover, nonadjacent dependencies are systematically ordered and require simultaneous processing of different structures to link these co-dependencies. Our nonadjacent dependency example above involves a nested center-embedded dependency, which are generally called as being recursive. The dependencies are embedded within one another as in the structure $a^1a^2a^3b^3b^2b^1 \ldots a^{n-1}\, b^{n-1}$, where a^{n-1} is the unit to be linked to b^{n-1} (Berwick et al. 2011). In contrast to this context-free grammar version, a finite-state form has the structure $a^1b^1a^2b^2a^3b^3 \ldots (ab)^{n-1}$, in which iterations are appended to the end of a pair.[10]

In examining acoustic pattern recognition in European starlings, Gentner et al. (2006) evaluated the syntactic recursion hypothesis that the capacity of self-embedding is unique to the human language faculty (Hauser et al. 2002). They conclude that the use of syntactic recursion is not unique to humans as their trained starlings have shown the ability to recognize these recursive patterns (Fig. 4.7).

However, Berwick and colleagues emphasize that the patterns tested do not match those sentence structures used by humans. While European starlings seemed to be able to recognize nesting after training, which needs to be differentiated from natural settings, they did not provide evidence of the bird's ability to recognize dependencies of the kind that pairs specific *as* with specific *bs*. The starlings tested above recognized an equal number of *as* and *bs* in terms of rattle and warble sound classes (rattle–warble patterns), but it has not been examined whether these rattle–warble patterns were properly paired off. Thus, at this point there is lack of evidence that songbirds or any other nonhuman species are able to use a strict context-free grammar. Birdsongs seem to be able to generate hierarchical structures, which is one basic principle of human syntax. For example, acoustic features of rattle or warble are not used to classify the rattle–warble sequence as a new unit, which in turn could be then used in more complex structures. However, some songbirds such as Bengalese finches seemed to be able to perceive 3–4 notes as a single unit (Suge and Okanoya 2009; see also Berwick et al. 2011). In this study, Fodor et al.'s classical psycholinguistic click experiment was applied, in which human subjects were presented with a click in the middle of a phrase (e.g., *ate the apple*). The participants

[10] It needs to be considered that a restricted phrase structure grammar can be approximated by a more elaborated finite-state grammar (e.g., memorized list of exemplars). Thus, it is important to use appropriate control trials such as the distinction between ungrammatical and grammatical strings).

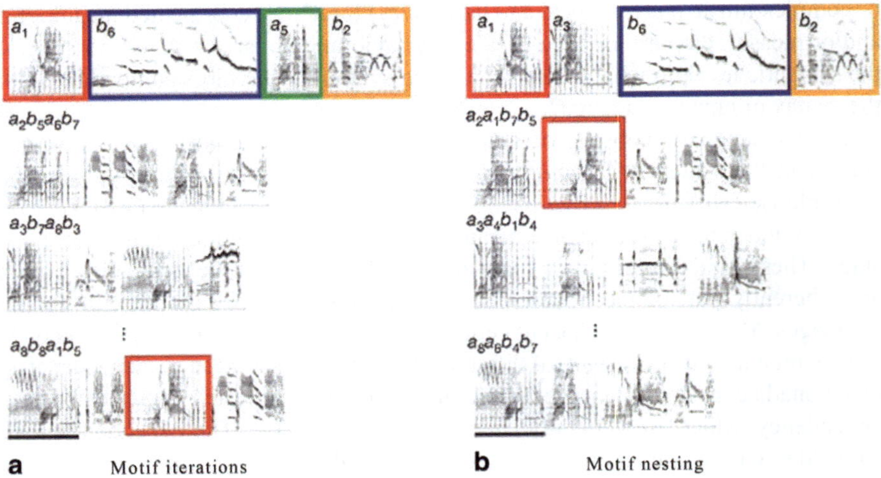

Fig. 4.7 Conditioned European starlings' sonograms of (**a**) rattle–warble iterations and (**b**) rattle–warble nesting. (Adapted and modified, Gentner et al. 2006; © Nature Publishing Group)

tend to report that the click occurred at the beginning or end of the phrase. In using this click protocol, Suge and Okanoya reported that the Bengalese finches responded in a similar fashion, that is, they perceived the click at the *c* or *e* of a *cde* unit. The Bengalese finches not only perceive such units but can also reproduce them. Thus, it appears that they are able to combine notes into units much like humans use words to combine them to noun or verb phrases and to use these units somewhere else. However, in contrast to human syntax, Bengalese finches seem to lack the ability of dependent nesting to manipulate these combined notes. In this vein, it has been demonstrated that cotton-top tamarins (Saguinus oedipus) are able to parse synthetic stimuli sequences generated by finite-state grammars, but not those generated by a phrase structure that implies a simple recursively hierarchical structure (Fitch and Hauser 2004). It is characteristic for human syntax that not a specific order of individual words is acquired but the specific order of word classes, also called phrase structures. Each single word can be inserted into a syntactic structure that provides a slot for this particular word class.

Also, new words can be created or as in the case of metaphors or idioms, for instance, a sequence of different word classes can be treated as a unit that the syntactic structure does not map onto the semantic structure in a combinatory fashion. To understand the meaning of *It rains cats and dogs,* the phrase needs to be processed as a single unit, as a string of words, and does not require combinatory computations between the individual words. Such kind of linguistic parsing flexibility to express nuances of different meanings seems to be unique to humans. The lack of a syntactic–semantic interface in nonhuman species (e.g., bonobos or bottle-nose dolphins) seems also to be reflected in the perception of verb argument structures (Kako 1999).

Table 4.1 Priming study to examine the effect of phonetic (pho) and semantic (sem) attributes in female Diana monkeys. (Zuberbühler et al. 1999)

	Prime	Probe	Attributes	Response
Baseline	Eagle shrieks	Eagle shrieks	+pho/+sem	Weak
	Leopard growls	Leopard growls	+pho/+sem	Weak
Test	Monkey eagle alarm	Eagle shrieks	−pho/+sem	Weak
	Monkey leopard alarm	Leopard growls	−pho/+sem	Weak
Control	Monkey leopard alarm	Eagle shrieks	−pho/−sem	Strong
	Monkey eagle alarm	Leopard growls	−pho/−sem	Strong

If a nonhuman animal uses arguments for communication, it would need to know the number and type of participants (arguments) involved in an event labeled by the action, the verb (Jackendoff 1987). The empirical evidence is weak. One reason is that more suitable tests need to be created to adjust to the specific situation of an animal's environment in a natural setting.[11] By sharing Bickerton's (1990) view, we assume that sensory experiences are stored as basic concepts, which consists of verb arguments. The first steps of language, which express meanings, but not vocalization *per se* may have involved labeling these basic concepts. Thus, some universal linguistic categories such as noun or verb may have evolved from these initial verb arguments. The rise of cognition in the genus Homo presumably resulted from the development of verb argument system, that is, of recognizing the relationship between different discrete objects. But let us return in this context to the results of some animal studies conducted to find out more about conceptualization in nonhuman primates, which may represent the evolutionary prerequisite of the human BDL.

It is obvious that many animals vocalize to inform their buddies of an event, which is related to food, danger, or territory. For instance, vervet monkeys have distinct calls for different predators such as leopards, eagles, and snakes and they respond to these calls although they have not seen these predators. A similar referential behavior has been studied in other species such as chimpanzees, bonobos, bottle-nose dolphins, sea lions, and parrots. However, the question is whether responses to specific vocalizations are solely perceptually driven or whether an internalized experience, a concept activated by the call, triggers the response. Current research indicates that at least our closest living relatives—genus *Pan,* monkeys, and other mammals—seem to make use of acoustic signals but also of concepts. To provide an example, in Zuberbühler et al. (1999) study, Diana monkeys' responses to various stimulus sequences were recorded (Table 4.1). In three different conditions (baseline, test, and control) the prime was varied while the probe represented eagle shrieks or leopard growls. In the baseline condition, the response of female Diana monkeys were weak to the probe as primes and probes were identical, that is, they shared the same phonetic and semantic attributes. In the test condition, the

[11] Although the distinction between a natural and a trained setting is important, overall it may be less significant for understanding the plasticity of an animal's cognitive capacity. Most human cognitive skills require training to be recognized as such. One might assume that even in animals cognitive capacities may change or increase as result of continuous training across many generations.

prime matched the probe at the semantic but not at the phonetic level, that is, the male Diana monkeys' alarm call for an eagle or leopard respectively matched the actual call of the predator. Zuberbühler and colleagues assumed that if the female monkey would respond strongly to the probe they would not associate the predators' call with the alarm call at the semantic level as both vocalizations do not share the same acoustic attributes. However, since the female monkeys' responses were attenuated, it was concluded that they indeed considered semantic information. The control condition confirmed this outcome as both the prime and probe pairs were incongruent with respect to phonetic and semantic attributes.

We might be not far from tracing back our language abilities to mental representations found in other primates and mammals. Language can be considered as a descendent of ancient cognitive abilities, whereas some subcomponents such as auditory–vocal learning may share properties across different species.

Finally, let us have a closer look at the complex vocalization skills of humpback whales (Megaptera novaeangliae), which rely more on singing rather than on sight or smell. One reason might be that sounds travel four times faster in water than in air. The highly aesthetic songs consist of a repeated series of sounds, apparently without any breaks. Each series holds for a period of 7–30 min (Payne and McVay 1971). The songs of humpback whales show similarity to some birdsongs found in Panamanian yellow-rumped cacique (Cacicus cela vitellinus) and village indigo bird (Vidua chalybeata). The similarity is that a song is group-specific and all members of the group have a similar repertoire, which successively changes. This phenomenon has been also called as a cultural evolution, whereas modulated songs are passed between individuals. In the case of humpback whales, males sing apparently to attract a female partner and/or to compete with other males. Noad et al. (2000) reported such a kind of cultural evolution: Within a period of 2 years, the song of humpback whales off the east coast of Australia changed significantly as 2 whales out of 82 introduced a new song. All male whales adopted the new hit in less than 2 years by learning these new vocal patterns. But of course, it is difficult to research the reasons or motivations that trigger this cultural development. It is known that humpback whales often modify their songs; however, no evidence has been reported that these modifications have different purposes or express different meanings. The analyses of the humpback whale songs point to a hierarchical, recursive structure, but in contrast to human language, there is no conceptual intent associated with these songs (Payne and McVay 1971; Suzuki et al. 2006). If the assumption of recursive patterns in humpback whale songs can be confirmed, the recursion hypothesis postulated to differentiate between animal communication systems and human language (Hauser et al. 2002) needs to be rejected, or at least, refined.

The evidence reported here so far shows that the communication system of each single species is unique in its characteristics at different sensory-specific levels, and if applicable, at different conceptual and linguistic levels. Future research will certainly reveal further overlaps between human and animal communication, but further differences will also shed light on the environmental and neurobiological criteria scaffolding the evolution of species-specific communication systems.

4.2 Proto-Cognition

One of the fundamental ideas presented here implies that the birth of language has its roots in the evolution of something what we might call "proto-cognition." We assume that ancestors of modern humans such as Homo habilis or H. erectus were equipped with a cognitive capacity, which was less developed and complex as compared to (ancestral or modern) H. sapiens, but more complex than the cognitive capacity of species classified as belonging to the genus Australopithecus. Thus, we assume that our ancestors of the genus Homo not only did communicate but were creators of tools and of cultural forms at the proto-cognitive level. To some extent, we are able to map homologous cortical areas between monkeys, chimpanzees, and modern humans. Also, tremendous Sisyphean work of paleoanthropologists reveals the anatomy of some of our ancestors in the hominid lineage and what kind of tools they manufactured and used. However, precise information about their social and cultural behavior and about their language and beliefs remains largely speculative. Some important clues, however, seem to contribute to solving this complex puzzle. One is related to the systematic increase of the brain volume in the human lineage. We hypothesize here that the increase of the brain size correlates with improved cognitive and linguistic capacities. It is certainly true that brain size *per se* cannot account for the ability to develop certain cognitive skills as specific neural wiring is required. However, in considering the complexity of cognitive and linguistic structures used by modern humans, it is quite reasonable to assume that cortical volume is related to multimodal fiber projections. Furthermore, we believe that proto-speech did not evolve isolated from other cognitive functions, but represents a part of the aforementioned proto-cognition. For instance, the human auditory–speech system did not evolve independent of nonvocal cognition but evolved in an interactive fashion.

A good example is the larynx position in different lineages. Chimpanzees, for example, live in groups of ca. 15–20 individuals and hoot, scream, grunt, and drum on hollow trees (in contrast to bonobos, who produce a high pitch). Thereby, wild chimpanzees relate meanings to their calls and combine these calls, but seem not to freely reorder or recombine speech motor commands (Crockford and Boesch 2005; Slocombe and Zuberbühler 2007). They use a relatively rich and sophisticated variation of a single call, probably to communicate within the low visible forest environment. For example, chimpanzees produce different grunts for different food amounts and different barks for different predator threats. However, it contrast to humans the range of vocalization is limited due to anatomical features (Lieberman 1968). In (modern) human infants as well as in adult chimpanzees (and in most but not all mammals), the larynx is located high in the throat, at the level of the C2–C3 vertebrae. As the human child grows, the larynx descends to the level of the C3–C6 vertebrae. This ontogenetic development probably mimics phylogenetic developments and this process is necessary that the child has access to the full phonetic range used in speech. However, a descended larynx is not necessarily a unique feature of the human vocal tract. X-rays of the vocal tracts of living nonhumans indicate in contrast to postmortem examinations that the larynx position is more flexible and descended as previously assumed (Fitch 2000, 2002). Some aquatic mammals (e.g., sea lions, walruses) but also terrestrial animals (e.g., dogs, pigs, goats, monkeys, deer, and young chimps) have a permanent or temporarily descended larynx (Hauser and Fitch 2003; Nishimura et al. 2003; McElligott et al. 2006). The Kyoto primate research group concluded that the condition for vocalization due to a descent of the larynx must have been partially evolved during the ape evolution (common ancestor of chimpanzees and humans), but not at once during human evolution. They scanned with structural MRI the brain and neck of three living chimpanzees and found that the larynges descend during infancy. A rapid descent of the laryngeal skeleton relative to the lingual bone (hyoid) was the reason, whereas the hyoid itself did not descend. Thus, it seems that evolution prepared human vocalization in two different phases for reasons that may be not necessarily related to vocalization but to adapt to different swallowing and breathing needs caused by changes in ape diet and increase in body size (e.g., Fleagle 1999; Nishimura, 2003). As mentioned before, how the larynx position changed during the hominid evolution remains an open question as fossil evidence allows only limited conclusions. But we know that the vocal tract of hominids gradually evolved and did not appear as the result of sudden mutation.

But what does the larynx position tell us about the cognitive capacity to speak a language? Indeed, a small set of phonetic variations does not necessarily rule out the development of a BDL. For instance, one might think of a language, which is less complex, a mini-language, which makes use of a small range of phonological, syntactic, and semantic rules and lexical units. This idea is to some extent compatible with the assumption of a proto-language. It has been argued, for instance, that a proto-language of our ancestors may have lacked a fully developed syntax, inflections, auxiliaries, or grammatical words (Bickerton 2009). Thus, the BDL would have been less elaborated and complex as in modern humans. Comparative research does not point to the assumption that the gap between proto-language and language has been filled by sudden mutations. Instead, modern language is the result of gradual emergence of cognition, physiological constraints, and biological disposition, and we can assume that this process involves multiple intermediate stages. Two classes of evidence support this assumption: the development of tool use and the increase of the brain size in the human lineage. Let us first discuss how cortical structures incrementally increased during human evolution.

We do not exactly know when the transitions occurred from modality-specific to cross-modal neural projections, and a precise picture about the gradual development toward language remains speculative. As mentioned before, we hypothesize that the significant increase of the brain size in the species H. erectus may indicate those neurobiological changes, which are required for developing a proto-language. The brain size was increased over H. erectus' cousins H. habilis, ranging between 850 and 1100 cc. As brain size alone provides only indirect information about the neural structure supporting specific cognitive functions, it cannot be ruled out that Homo rudolfensis (ca. 775 cc) and H. habilis (ca. 600 cc) are also candidates for proto-language. However, H. erectus seems to be the prime candidate for proto-

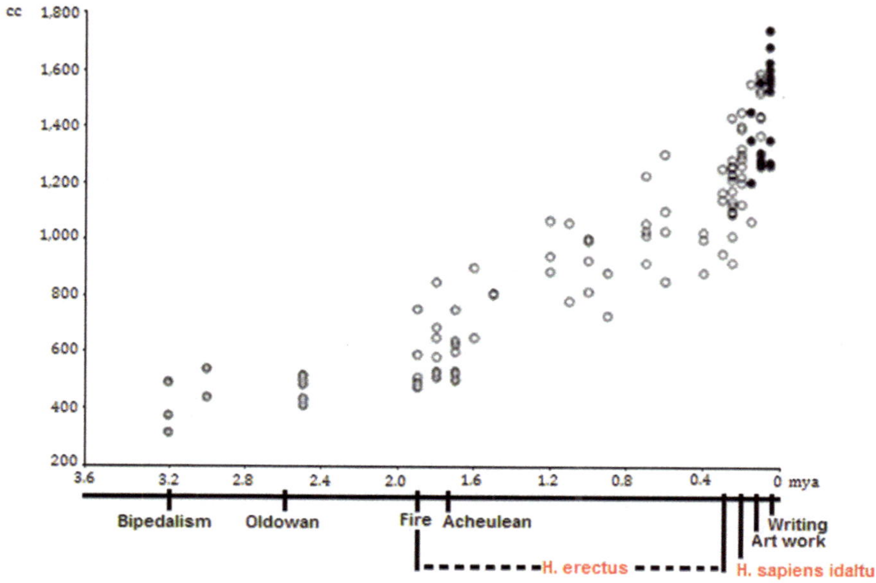

Fig. 4.8 Cranial capacity and significant behavioral events of human evolution: ∘ Australopithecus, Paranthropus; • Neanderthals; o all other Homo; ☆ H. sapiens; □ Gorilla; Δ Chimpanzee. (Adapted and modified, Flinn et al. 2005; © Elsevier Limited)

language. Besides the significant increase of cranial capacity, this species lived longer than any other hominid species. About 2–1 mya, early Homo ergaster migrated outside the aforementioned valley system into the surrounding area of Africa, and then from there into Central and Eastern Asia—evolving along the way into H. erectus. Apparently, the geographic distances reduced substantially the gene flow in the total hominid population, making regional subspecies (such as eastern H. erectus) more likely to appear. Besides climate change and fluctuating resources, hominid tools, diet, and child rearing now emerged as factors that could increasingly affect human group organization and patterns of dispersal. It should be considered that the specified brain sizes of our hominid ancestors are estimates from endocasts, which are made from the inside of reassembled fossil skulls. This method cannot capture convolutions of fissures. Thus, it might be that our brain is substantially more convoluted than the brain of a H. erectus. Also, the brain of an H. erectus could be slightly larger when adjusted for body size. In contrast to all previous hominoids, H. erectus had a large body size, which correlated, thus, with his relatively large brain. Figure 4.8 shows that between 2–0.7 mya, the brain size doubled in this species.

The exchange of communicative signals and the development of symbolic capacity turned out to be a highly beneficial behavior, which changed epigenetic processes. In particular, the frontal lobe increased in Australopithecus africanus and subsequent hominoids. Indeed, some argue that the frontal lobe is the key for the development of social behavior, planning, and language (Deacon 1997). The ability to make tools appeared 2.4 mya, and thus, frontal lobe expansion may have already occurred. Furthermore, we believe that proto-language may have already developed, which combined two or more signs by using a basic syntax. The skull of modern humans has a more rounded shape as compared to earlier species. It might indicate an expansion of the middle parietal lobe, which plays an important role in integrating different kinds of sensory information, in particular visuo-spatial information. The parietal expansion in modern humans might be related to the refinement of tool cultures and abstract and complex thinking around 100–50 thousand years ago (kya). Whether the expansion of the parietal lobe reflects another milestone toward modern language competencies remains to be seen. Most interesting is that the frontal and parietal lobe expansion seemed to have occurred in parallel but reflect different evolutionary stages. The evolutionary account supports the idea of a modularized architecture of the brain, but which operates in a highly interactive manner. The idea of evolving cognitive modules of language skills is another indicator for the assumption that modern language gradually developed and has its roots in communication forms (e.g., a combination of sounds and signs) fulfilling only portions of the role, which today's natural languages achieve.

Doubtlessly, brain size and encephalization quotient (EQ) substantially increased during human evolution.[12] A more recent study proposes that the EQ of the prefrontal cortex, which is associated with social and working memory competencies, modestly increased by ca. 10%. Thereby, the increase of the human neocortex results in relatively more interconnections among neurons, axons, and white matter compared to the absolute number of neurons such as cell bodies and gray matter (Holloway 2002; Zhang and Sejnowski 2000). Thus, the evolutionary changes may particularly involve an increase of cross-modal interconnections between different brain regions. Again, we can predict that a higher EQ is associated with greater cortical folding (gyrification), which in turn may lead to increased neural modularization of cognitive functions. The ACC that represents a collar-form around the corpus callosum (BAs 24, 32, and 33) regulates in addition to blood pressure and heart rate various cognitive functions such as social cognition, decision making, empathy, and emotion. Another part of the prefrontal cortex has been reorganized during hominid evolution. According to Semendeferi et al. (1998), BA 13 is in modern humans about half of the size when the total brain size is considered. They assume that neighbored regions (BAs 11 and 13) became specialized in humans for processing socially-related information. Already, Holloway and de La Costelareymondie (1982) reported specific asymmetric cortical regions among both hemispheres in H. erectus and H. sapiens when analyzing the shape of the endocasts. BAs 17 and 18 of the left occipital lobe are smaller and the left parietal area and the right frontal cortex are larger compared to the contralateral homologs. Similar, in human right-handers the dorsolateral prefrontal cortex and the frontal pole (BAs 46

[12] EQ is a measure of relative brain size. It reflects the ratio between actual brain mass and predicted brain mass of an organism of a given body size.

and 10) seem to be larger in the right hemisphere, but this asymmetry was not found in chimpanzees (Zilles et al. 1996).

One might argue that the asymmetric cortical evolution of right prefrontal areas is related to the cognitive development of self-awareness and social cognition. This cortical distribution may be also related to the evolution of a "theory of mind" (ToM), and has been supported by other studies (e.g., Saxe and Powell 2006). The concept of a ToM refers to the cognitive ability to understand own mental states and of others, that is, to be able to put your feet in other people's shoes. There is no doubt that language is closely tied to self-awareness by generating a mental scheme. It enables us to communicate with others about events removed from space and/or time. The speaker can use symbolic units to refer to him-/herself, other persons, objects, events, etc. to reflect on his/her and others' mental states, which serve a social function. In contrast, mathematics can be considered as a special case of linguistic competencies, in which symbolic units are manipulated according to a logical system: "All mathematics is symbolic logic." (Russel 1903). Thus, it is to assume that the development of the BDL is directly related to an increase of attention and working memory capacities. Since we believe that linguistic and cognitive skills co-evolved, we assume that working memory capacities and other cognitive abilities relevant for language processing (e.g., increased attention to sound patterns) sustained each other. The evolution of language may have, therefore, significantly contributed to the development of awareness. One may, therefore, argue the nonlinear increase of the brain size as compared to the body size may be associated to the co-evolution of language and cognition. Moreover, we assume that climate conditions significantly influenced social behavior among ancient H. sapiens. Harsh living conditions in east Africa such as drought may have led to increased migration pressure around 200 kya. A long glacial stage (MIS6: marine isotope stage 6), which deteriorated the conditions in Africa around that time, and the human population plummeted to a few hundred (Marean 2010). This view goes in line with the bottleneck model: As indicated by genetic studies, we all are descending from a small population in Africa, which lived during MIS6. Thus, the significant rise of cognition may have emerged from the struggle to survive. Thereby, the improvement of communication skills in form of language may have played a significant role.

The development of tool making and usage is one important class of evidence for the cognitive abilities of our ancestors. The species Australopithecus, for instance, might have used, 4–1 mya, unshaped tools for basic purposes such as butchering animal remains. Otherwise, it is assumed that the australopith species was not more adept to tools than modern apes such as chimpanzees or gorillas. Gorillas, for example, open nuts with stones or using long sticks to dig for terminates and chimpanzees use sometimes spears, crack nuts, clip leaves, and fish and dip for ants by using a wand (Alp 1993). It seems that only certain species of Australopithecus were able to use unshaped tools in a different form than nonhuman primates. The Stone Age is a prehistory cultural period roughly lasting 3 million years up to 2000 BC. It is typically divided into Paleolithic, Mesolithic, and Neolithic periods and reflects an increased sophistication of tool manufacturing and use. The development of tool making has been strongly influenced by climate changes. The northern latitudes

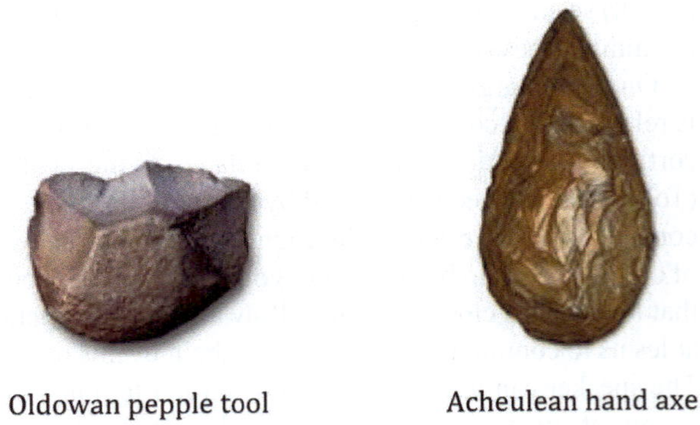

Fig. 4.9 Examples of an Oldowan pebble tool and an Acheulean hand axe. The Oldowan tools were used by A. garhi, H. habilis, and early H. erectus, and Archeulean tools by H. erectus, H. heidelbergensis, Homo neanderthalis, and H. sapiens idaltu. (Adapted and modified, Wikipedia)

were subject to four successive advances and retreats of ice sheets. The first stone technology known and used during the Lower Paleolithic period by Australopithecus garhi is called Oldowan (Mode 1: 2.6–1.8 mya)[13]. These tools consist of core and flakes, whereas the latter ones were produced by hitting the stone core with a hammer stone to get flakes. Stone cores and flakes were used to cut, brash, and scrape. Besides A. garhi (2.6–2.3 mya; brain size ca. 450 cc), it is assumed that H. habilis (2.3–1.4 mya; ca. 600 cc) and H. ergaster/early H. erectus (1.8–1.3 mya; early fossils: 700–900 cc; later fossils: 900–1100 cc) seemed to have made use of the Mode 1 technology but further refined it, 1.7 mya, toward the Acheulean technology (Mode 2: 1.7–0.1 mya).14 The late Mode 2 tools were also use by anatomically modern humans (AMH) such as H. sapiens idaltu (dated 160–150 kya; 1,450 cc; Neanderthal species appeared 600–350 kya; mean 1450 cc) and Homo heidelbergensis (600–400 kya; mean 1350 cc, equivalent to AMH), the ancestor of Neanderthals and AMH (Fig. 4.9).

Again, the Acheulean was replaced in Europe by the Mousterian industry (Mode 3: 300–30 kya), a lithic technology mostly used by Neanderthals. The tools of both industries were often manufactured with the Levallois technique, which involves flaking pieces off the edge of a large piece of flint until it looks like a turtle shell. Again, by hitting this core of the flint, flakes with a distinctive plano-convex profile are separated, while its edges have been previously sharpened by trimming the core. Again, different tool industries developed in the Upper Paleolithic epoch (Mode 4) as not only flint was used but also bone, ivory, and antler. Hand axes and flake tools were replaced by more complex and more specialized tools such as needles and thread, skin clothing, hafted stone and bone tools, harpoon, spear thrower, and special fishing equipment. Without any doubt, during the transition from Middle to Upper Paleolithic (ca. 50–30 kya) art work seemed to have appeared for the first time along with agriculture and civilization. During the Mesolithic era (11 kya),

---

[13] The term "Oldowan" (pebble tools) refers to Olduvai Gorge in Tanzania, where the first stone tools were found.

[14] The term "Acheulean" refers to a region in Amiens, northern France, where hand-axes were found in the middle of the nineteenth century.

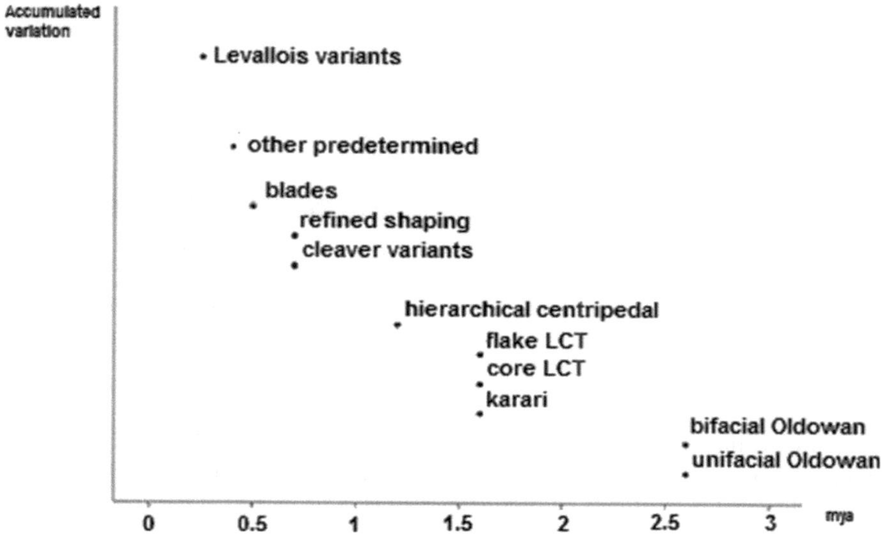

Fig. 4.10 Accumulation of Paleolithic tool variations. Each innovation represents an increment of 1 and some points (e.g., cleaver variants) correspond to more than one innovation. LCT, large cutting tool. (Adapted and modified, Stout 2011; © Elsevier Limited)

glaciers retreat (mostly north of 50° N in Eurasia and North America) and hunting of herd animals was replaced by exploitation of forest resources. The Neolithic epoch is part of the late Stone Age and preceded the Bronze Age and followed the Paleolithic period. The time period depends on geographic regions (e.g., China: 10,000–2,000 years BC; northern India, 8,000–4,500 years BC; Eastern Mediterranean ca. 10,000–3,300 years BC, and eastern and northern Europe ca. 3,000–1,800 years BC). With the exception of H. sapiens, all other human species were extinct. People still used stone tools, but domesticated plants and animals and lived in permanent settlements or villages.

It is difficult to correlate directly the spatial ability of tool making such as symmetry of hand axes with communicative behavior. However, the motor areas for fine movements are located in the same brain region, which also controls speech. Again, some neuroimaging studies indicate complex motor patterns for stone tool knapping as well as for speech in Broca's region (Stout 2008, 2011). In particular, an increase of cortical activity was found for the Acheulean mode as compared to the Oldowan mode. Thus, the making of relatively refined tools at around 1.7 mya (Mode 2: Acheulean tools) may indicate a more refined articulation.

The development of tool making in the Lower Paleolithic indicates a cumulative cognitive – cultural evolution and seems to accelerate much like the evolution of cortical structures (Fig. 4.10). Stout (2011) developed a system of action hierarchies with increasing complexity from Oldowan flake production to late Acheulean shaping. In particular, the ability to generate hierarchical structures is a prominent

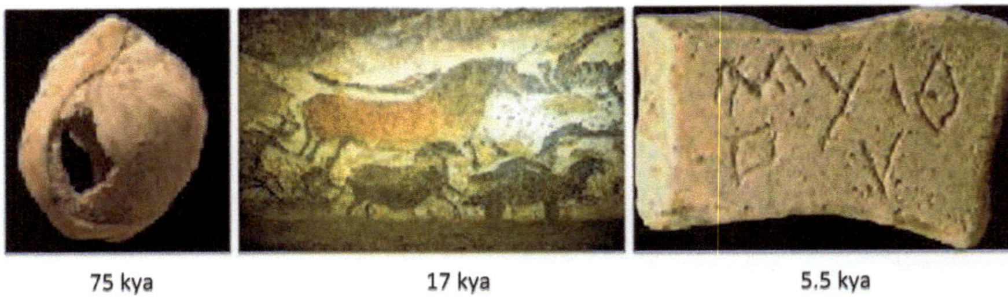

Fig. 4.11 One of the oldest symbolic art forms and symbols found in prehistory. Left: Shells discovered in the Blombos cave, South Africa. Middle: Cave paintings in the Lascaux caves of southwestern France. Right: Symbolic markings probably evolved later into the Indus script were found on fragments of pottery in Pakistan. (Adapted and modified, Wikipedia)

principle of linguistic structure generation beyond the word level. It is, therefore, plausible to assume that H. erectus was able to form basic hierarchical structures across different modalities, possibly including singing and dancing.

Whether neural activation patterns transferred from tool making and/or even from other modalities to speech or whether the refinement of articulation was an independent or parallel evolving process, remains to be seen. But it appears that the neurobiological capacity for tool making in general and for speech was already available during Mode 1 by H. habilis and/or H. erectus. A new cognitive–cultural level based on the previous one can be characterized by refinement, modifications, extensions, and changes. In general, the cognitive–cultural evolution progresses independent of biological evolution and has its "eigen-dynamics," although we do not deny complex interactions between both evolving processes at different levels. The development of sophisticated and complex cultures from computer technology, machineries to architecture takes thousands of generations but is not the product of biological mutations. In contrast, if we ignore the use of proto-language in ancient history at this point, human language is an art form that changes along with cultural conditions.

There is unequivocal evidence for the creation of human art after 50 kya, that is, after the emergence of modern humans. However, older discoveries such as color symbolism may indicate the use of art elements. In the Blombos cave (300 km east of Cape Town) a processing workshop was found, where a liquefied ochre-rich mixture was produced and stored in two shells (abalone) about 100 kya (Henshilwoo et al. 2011). At the same site, 41 shells were found earlier in the same cave, which have marks of red ochre and holes in symmetric positions. The shells presumably brought from rivers a dozen miles away are dated to be 75,000 years old and may be considered as one of the oldest jewelries (Fig. 4.11).

In the Middle Paleolithic (100–50 kya), only a few art products were discovered such as the shells of the Blombos cave. Undisputed art work is from Upper Paleolithic (40–10 kya), and includes rock paintings or carvings and portable sculptures. They were found on different continents in Europe, Africa, the Middle East, Asia, and Australia. By 12 kya, one of the oldest art forms found in North America is the painting of a Cooper bison skull, and in South America, rock paintings in the Toquepala caves in Peru. Most prehistory cave paintings tell a story about experienced events and do not show single objects. One of the first documented maps (ca. 10 kya) seems to show dwellings along a river and was found in Mezhirich, Ukraine. Again, one of the first musical instruments is a bone flute discovered in China (9 kya). In general, paintings and drawings seem to predate other symbolic art products including beads, sculptures, figurines, music instruments, and textiles. But in particular, writing seems to have systematically developed in at least three different locations (ca. 5,500 years ago): Egypt, Mesopotamia, and Harappa (Indus script). However, Indus scripts are difficult to decode as these languages died. In contrast, the famous Rosetta Stone (ca. 2,200 years ago) is a large slab of black basalt inscribing a decree in ancient Egyptian hieroglyphs, Demotic script and lowest Ancient Greek, which allowed to decipher the first time the Egyptian hieroglyphs.

Symbolic expressions indicate abstract thinking and developed several hundred thousand years earlier than the development of writing systems. If we consider Acheulean tool making as an indicator for the beginning of abstract thinking in the form of planned actions, the temporal gap between symbolism and writing might cover a period of up to 1 million years. The evidence discussed so far clearly demonstrates that bipedalism preceded brain growth, which in turn parallels the cognitive–cultural development of tool making. But when did our ancestors use symbolic means to communicate? Here, we must differentiate between a language-ready brain (BDL) and the actual use of communication forms.

References

Aboitiz, F. (2012). Gestures, vocalizations, and memory in language origins. *Frontiers in Evolutionary Neuroscience, 4*(2).

Alp, R. (1993). Meat eating and ant dipping by wild chimpanzees in Sierra Leone. *Primates, 34*(4), 463–468.

Arbib, M. A. (2005). From monkey-like action recognition to human language: an evolutionary framework for neurolinguistics. *The Behavioral and Brain Sciences, 28*(2), 105–124; discussion 125–167.

Arriaga, G., & Jarvis, E. D. (2013). Mouse vocal communication system: Are ultrasounds learned or innate? *Brain and Language, 124*(1), 96–116.

Arriaga, G., Zhou, E. P., & Jarvis, E. D. (2012). Of mice, birds, and men: The mouse ultrasonic song system has some features similar to humans and song-learning birds. *PLoS One, 7*(10), e46610.

Berwick, R. C., Okanoya, K., Beckers, G. J. L., & Bolhuis, J. J. (2011). Songs to syntax: The linguistics of birdsong. *Trends in Cognitive Sciences, 15*(3), 113–121.

Bickerton, D. (1990). *Language and species*. Chicago: The University of Chicago Press.

Bickerton, D. (2009). *Adam's tongue: How humans made language, how language made humans*. New York: Hill and Wang.

Bolhuis, J. J., Okanoya, K., & Scharff, C. (2010). Twitter evolution: Converging mechanisms in birdsong and human speech. *Nature Reviews Neuroscience, 11*(11), 747–759.

Bosman, C., García, R., & Aboitiz, F. (2004). FOXP2 and the language working-memory system. *Trends in Cognitive Sciences, 8*(6), 251–252.

Broadfield, D. C., Holloway, R. L., Mowbray, K., Silvers, A., Yuan, M. S., Márquez, S. (2001). Endocast of Sambungmacan 3 (Sm 3): A new *Homo erectus* from Indonesia. *Anatomical Record, 262*(4), 369–379.

Brodmann, K. (1909). *Vergleichende Lokalisationslehre der Grosshirnrinde*. Leipzig: Johann Ambrosius Bart.

Buchsbaum, B. R., Olsen, R. K., Koch, P., & Berman, K. F. (2005). Human dorsal and ventral auditory streams subserve rehearsal-based and echoic processes during verbal working memory. *Neuron, 48*(4), 687–697.

Cantalupo, C., & Hopkins, W. D. (2001). Asymmetric Broca's area in great apes. *Nature, 414*(6863), 505.

Catani, M., Jones, D. K., & ffytche, D. H. (2005). Perisylvian language networks of the human brain. *Annals of Neurology, 57*(1), 8–16.

Corballis, M. C. (2003). From mouth to hand: Gesture, speech and the evolution of right-handedness. *The Behavioral and Brain Sciences, 26*(2), 199–208.

Crockford, C., & Boesch, C. (2005). Call combinations in wild chimpanzees. *Behaviour, 142*, 397–421.

Crockford, C., Herbinger, I., Vigilant, L., & Boesch, C. (2004). Wild chimpanzees produce group-specific calls: A case for vocal learning? *Ethology, 110*(3), 221–243.

Darwin, C. (1859). *On the origin of species by means of natural selection, or the preservation of favoured races in the struggle for life*. London: John Murray.

Dax, M. (1836). Lésions de la moitié gauche de l'encéphale coïncident avec l'oubli des signes de la pensée. *Bulletin hebdomadaire de médecine et de chirurgie, 2me série, 2*, 259–262.

Deacon, T. W. (1992). Cortical connections of the inferior arcuate sulcus cortex in the macaque brain. *Brain Research, 573*, 8–26.

Deacon, T. W. (1997) *The symbolic species: The coevolution of language and the brain*. New York: Norton.

Dick, A. S., & Tremblay, P. (2012). Beyond the arcuate fasciculus: Consensus and controversy in the connectional anatomy of language. *Brain, 135*(12), 3529–3550.

Falk, D. (2007). Evolution of the primate brain. In W. Henke & I. Tattersall (Eds.), *Handbook of palaeoanthropology, Vol. 2: Primate evolution and human origins*. Berlin: Springer.

Falk, D. (2007). Constraints on brain size: The radiator hypothesis. In J. H. Kaas (Ed.), *The evolution of nervous systems* (pp. 347–354). Oxford: Academic.

Fitch, W. T. (2000). The phonetic potential of nonhuman vocal tracts: Comparative cineradiographic observations of vocalizing animals. *Phonetica, 57*(2-4), 205–218.

Fitch, W. T. (2002). Comparative vocal production and the evolution of speech: Reinterpreting the descent of the larynx. In A. Wray (Ed.), *The transition to language* (pp. 21–45). Oxford: Oxford University Press.

Fitch, W. T., & Hauser, M. D. (2004). Computational constraints on syntactic processing in a nonhuman primate. *Science, 303*(5656), 377–380.

Fleagle, J. G. (1999). *Primate adaptation and evolution*. San Diego: Academic.

Flinn, M. V., Geary, D. C., & Ward, C. V. (2005). Ecological dominance, social competition, and coalitionary arms races: Why humans evolved extraordinary intelligence. *Evolution and Human Behavior, 26*, 10–46.

Frey, S., Campbell, J. S. W., Pike, G. B., & Petrides, M. (2008). Dissociating the human language pathways with high angular resolution diffusion fiber tractography. *The Journal of Neuroscience, 28*(45), 11435–11444.

Friederici, A. D., & Gierhan, S. M. (2013). The language network. *Current Opinion in Neurobiology, 23*(2), 250–254.

Galaburda, A. M., & Pandya, D. N. (1982). Role of architectonics and connections in the study of primate brain evolution. In E. Armstrong & D. Falk (Eds.), *Primate brain evolution: Methods and concepts* (pp. 203–216). New York: Plenum.

Gallese, V., Fadiga, L., Fogassi, L., & Rizzolatti, G. (1996). Action recognition in the premotor cortex. *Brain: A Journal of Neurology, 119*(2), 593–609.

Gannon, P. J., Holloway, R. L., Broadfield, D. C., & Braun, A. R. (1998). Asymmetry of chimpanzee planum temporale: Humanlike pattern of Wernicke's brain language area homolog. *Science, 279*(5348), 220–222.

Gentner, T. Q., Fenn, K. M., Margoliash, D., & Nusbaum, H. C. (2006). Recursive syntactic pattern learning by songbirds. *Nature, 440*(7088), 1204–1207.

Glasser, M. F., & Rilling, J. K. (2008). DTI tractography of the human brain's language pathways. *Cerebral Cortex, 18*(11), 2471–2482.

Hauser, M. D., & Fitch, W. T. (2003). What are the uniquely human components of the language faculty? In M. H. Christiansen & S. Kirby (Eds.), *Language evolution* (pp. 158–181). Oxford: University Press Scholarship.

Hauser, M. D., Chomsky, N., & Fitch, W. T. (2002). The faculty of language: What is it, who has it and how did it evolve? *Science, 298,* 1569–1579.

Henshilwoo, C. S., d'Errico, F., van Niekerk, K. L., Coquinot, Y., Jacobs, Z., Lauritzen, S. E., Menu, M., García-Moreno, R. (2011). 100,000-year-old ochre-processing workshop at Blombos Cave, South Africa. *Science, 334*(6053), 219–222.

Hickok, G., & Poeppel, D. (2004). Dorsal and ventral streams: A framework for understanding aspects of the functional anatomy of language. *Cognition, 92*(1-2), 67–99.

Hickok, G., & Poeppel, D. (2007). The cortical organization of speech processing. *Nature Reviews Neuroscience, 8*(5), 393–402.

Holloway, R. L. & de La Costelareymondie, M.C. (1982). Brain endocast asymmetry in pongids and hominids: Some preliminary findings on the paleontology of cerebral dominance. *American Journal of Physical Anthropology, 58*(1), 101–110.

Holloway, R. L. (2002). Brief communication: How much larger is the relative volume of area 10 of the prefrontal cortex in humans? *American Journal of Physical Anthropology, 118*(4), 339–401.

Hopkins, W. D., Marino, L., Rilling, J. K., & MacGregor, L. A. (1998). Planum temporale asymmetries in great apes as revealed by magnetic resonance imaging (MRI). *NeuroReport, 9,* 2913–2918.

Izumi, A. & Kojima, S. (2004). Matching vocalizations to vocalizing faces in a chimpanzee (*Pan troglodytes*). *Animal Cognition, 7*(3), 179–184.

Jackendoff, R. (1987). The status of thematic relations in linguistic theory. *Linguistic Inquiry, 18*(3), 369–411.

Jackson, W. J., Reite, M. L., & Buxton, D. F. (1969). The chimpanzee central nervous system: A comparative review. *Primates in Medicine, 4,* 1–51.

Jarvis, E. D. (2006). Evolution of vocal learning systems in birds and humans. In: J. Kass (Ed.), *Evolution of nervous systems, vol. 2* (pp. 213–228). Waltham: Academic.

Jürgens, U. (2003). From mouth to mouth and hand to hand: On language evolution. *Behavioral and Brain Sciences, 26*(2), 229–230.

Jürgens, U. (2009). The neural control of vocalization in mammals: A review. *Journal of Voice: Official Journal of the Voice Foundation, 23*(1), 1–10.

Kako, E. (1999). Elements of syntax in the systems of three language-trained animals. *Animal Learning & Behavior, 27*(1), 1–14.

Kelly, C. & Uddin, L. Q., Shehzad, Z., Margulies, D. S., Xavier Castellanos, F., Milham, M. P., & Petrides, M. (2010). Broca's region: Linking human brain functional connectivity data and non-human primate tracing anatomy studies. *European Journal of Neuroscience, 32,* 383–398.

Kohler, E., Keysers, C., Umiltà, M. A., Fogassi, L., Gallese, V., & Rizzolatti, G. (2002). Hearing Sounds, Understanding Actions: Action Representation in Mirror Neurons. *Science, 297*(5582), 846–848.

Lam, Y.-W., & Sherman, S. M. (2010). Functional organization of the somatosensory cortical layer 6 feedback to the thalamus. *Cerebral Cortex, 20*(1), 13–24.

Levréro, F., & Mathevon, N. (2013). Vocal signature in wild infant chimpanzees. *American Journal of Primatology, 75*(4), 324–332.

Liberman, A. M., & Mattingly, I. G. (1985). The motor theory of speech perception revised. *Cognition, 21*(1), 1–36.

Lichtheim, L. (1884). Ueber Aphasie. *Deutsches Archiv Für Klinische Medicin, 36,* 204–268.

Lieberman, P. (1968). Primate vocalizations and human linguistic ability. *Journal of the Acoustic Society of America, 44,* 1574–1584.
Marean, C. W. (2010). Pinnacle Point Cave 13B (Western Cape Province, South Africa) in context: The Cape Floral kingdom, shellfish, and modern human origins. *Journal of Human Evolution, 59*(3-4), 425–443.
Matelli, M., Luppino, G., & Rizzolatti, G. (1985). Patterns of cytochrome oxidase activity in the frontal agranular cortex of macaque monkey. *Behavioural Brain Research, 18,* 125–136.
McElligott, A.G., Birrer, M., & Vannoni, E. (2006). Retraction of the mobile descended larynx during groaning enables fallow bucks (*Dama dama*) to lower their formant frequencies. *Journal of Zoology, 270*(2), 340–345.
Nishimura, T. (2003). Comparative morphology of the hyo-laryngeal complex in anthropoids: Two steps in the evolution of the descent of the larynx. *Primates, 44,* 41–49.
Nishimura, T., Mikami, A., Suzuki, J., & Matsuzawa, T. (2003). Descent of the larynx in chimpanzee infants. *Proceedings of the National Academy of Sciences of the United States of America, 100*(12), 6930–6933.
Noad, M. J., Cato, D. H., Bryden, M. M., Jenner, M.-N., & Jenner, K. C. S. (2000). Cultural revolution in whale songs. *Nature, 408*(6812), 537.
Ogawa, S., Lee, T. M., Nayak, A. S., & Glynn, P. (1990). Oxygenation-sensitive contrast in magnetic resonance image of rodent brain at high magnetic fields. *Magnetic Resonance in Medicine: Official Journal of the Society of Magnetic Resonance in Medicine/Society of Magnetic Resonance in Medicine, 14*(1), 68–78.
Payne, R. S., & McVay, S. (1971). Songs of humpback whales. *Science, 173*(3997), 585–597.
Petkov, C. I. & Jarvis, E. D. (2012). Birds, primates, and spoken language origins: behavioral phenotypes and neurobiological substrates. *Frontiers in Evolutionary Neuroscience, 4*:12. doi: 10.3389/fnevo.2012.00012
Petrides, M., & Pandya, D. N. (2006). Efferent association pathways originating in the caudal prefrontal cortex in the macaque monkey. *The Journal of Comparative Neurology, 498*(2), 227–251.
Petrides, M., & Pandya, D. N. (2009). Distinct parietal and temporal pathways to the homologues of Broca's area in the monkey. *PLoS Biology, 7*(8), e1000170.
Pooley, R. A. (2005). AAPM/RSNA physics tutorial for residents: Fundamental physics of MR imaging. *Radiographics: A Review Publication of the Radiological Society of North America, Inc, 25*(4), 1087–1099.
Preuss, T. M., & Goldman-Rakic, P. S. (1991). Architectonics of the parietal and temporal association cortex in the strepsirhine primate *Galago* compared to the anthropoid primate *Macaca*. *The Journal of Comparative Neurology, 310*(4), 475–506.
Rauschecker, J. P. (2011). An expanded role for the dorsal auditory pathway in sensorimotor control and integration. *Hearing Research, 271,* 16–25.
Riede, T., Owren, M. J., & Arcadi, A. C. (2004). Nonlinear acoustics in pant hoots of common chimpanzees (*Pan troglodytes*): Frequency jumps, subharmonics, biphonation, and deterministic chaos. *American Journal of Primatology, 64*(3), 277–291.
Rilling, J. K., Glasser, M. F., Preuss, T. M., Ma, X., Zhao, T., Hu, X., & Behrens, T. E. J. (2008). The evolution of the arcuate fasciculus revealed with comparative DTI. *Nature Neuroscience, 11,* 426–428.
Rizzolatti, G., & Craighero, L. (2004). The mirror-neuron system. *Annual Review of Neuroscience, 27,* 169–192.
Rizzolatti, G., Fadiga, L., Gallese, V., & Fogassi, L. (1996). Premotor cortex and the recognition of motor actions. *Brain Research. Cognitive Brain Research, 3*(2), 131–141.
Rolheiser, T., Stamatakis, E. A., & Tyler, L. K. (2011). Dynamic processing in the human language system: Synergy between the arcuate fascicle and extreme capsule. *The Journal of Neuroscience, 31*(47), 16949–16957.
Russel, B. (1903). *The principles of mathematics*. Cambridge: Cambridge University Press.
Saur, D., Kreher B. W., Schnell S., Kümmerer D., Kellmeyer P., Vry M. S., Umarova R., Musso M., Glauche V., Abel S., Huber W., Rijntjes M., Hennig J, Weiller C. (2008). Ventral and dorsal pathways for language. *Proceedings of the National Academy of Science, 105*(46), 18035–18040.

Saxe, R. & Powell, L. J. (2006). It's the thought that counts: Specific brain regions for one component of theory of mind. *Psychology Science, 17*(8), 692–699.

Schmahmann, J. D., Pandya, D. N., Wang, R., Dai, G., D'Arceuil, H. E., Crespigny, A. J. de, & Wedeen, V. J. (2007). Association fibre pathways of the brain: Parallel observations from diffusion spectrum imaging and autoradiography. *Brain, 130*(3), 630–653.

Semendeferi, K., Armstrong, E., Schleicher, A., Zilles, K., and Van Hoesen, G. W. (1998). Limbic frontal cortex in hominoids: A comparative study of area 13. *American Journal of Physical Anthropology, 106,* 129–155.

Sherwood, C. C., Broadfield, D. C., Holloway, R. L., Gannon, P. J., & Hof, P. R. (2003) Variability of Broca's area homologue in African great apes: Implications for language evolution. *Anatomical Record A. Discoveries in Molecular Cellular Evolutionary Biology, 271*(2), 276–285.

Slocombe, K. E., & Zuberbühler, K. (2007). Chimpanzees modify recruitment screams as a function of audience composition. *Proceedings of the National Academy of Sciences of the United States of America, 104,* 17228–17233.

Stout, D. (2008). Technology and human brain evolution. *General Anthropology, 15,* 1–5.

Stout, D. (2011). Stone toolmaking and the evolution of human culture and cognition. *Philosophical Transactions of the Royal Society of London B, 366,* 1050–1059.

Suge, R., & Okanoya, K. (2009). Perceptual chunking in the self-produced songs of Bengalese finches (*Lonchura striata var. domestica*). *Animal Cognition, 13*(3), 515–523.

Suzuki, R., Buck, J. R., & Tyack, P. L. (2006). Information entropy of humpback whale songs. *The Journal of the Acoustical Society of America, 119*(3), 1849–1866.

Ungerleider, L. G., & Haxby, J. V. (1994). "What" and "where" in the human brain. *Current Opinion in Neurobiology, 4*(2), 157–165.

Wada, H., Sekino, M., Ohsaki, H., Hisatsune, T., Ikehira, H., & Kiyoshi, T. (2010). Prospect of high-field MRI. *IEEE Transactions on Applied Superconductivity, 20*(3), 115–122.

Wernicke, C. (1874). *Der aphasiche Symptomenkomplex. Eine psychologische Studie auf anayomischer Basis.* Breslau: Cohn & Weigert.

Wilson, S. M., Saygin, A. P., Sereno, M. I., & Iacoboni, M. (2004). Listening to speech activates motor areas involved in speech production. *Nature Neuroscience, 7,* 701–702.

Zhang, K. & Sejnowski, T. J. (2000). A universal scaling law between gray matter and white matter of cerebral cortex. *Proceedings of the National Academy of Sciences of the United States of America, 97*(10), 5621–5626.

Zilles, K., Dabringhaus, A., Geyer, S., Amunts, K., Qü, M., Schleicher, A., Gilissen, E., Schlaug, G., & Steinmetz, H. (1996). *Neuroscience Biobehavioral Reviews, 20*(4), 593–605.

Zuberbühler, K., Cheney, D. L., & Seyfarth, R. M. (1999). Conceptual semantics in a nonhuman primate. *Journal of Comparative Psychology, 113*(1), 33–42.

Part II
Paradigms

Chapter 5
The Human Language System

5.1 Biological Disposition

The biological disposition of language (BDL) has to be reflected in the anatomical structures of the human brain. Most neuroscientists would agree that the prime factor is the type of neural wiring but not volume that provides the base for our capacity to use language. However, it appears that language-related wiring would require a certain brain size as human's encephalization level is highest among the mammals. Encephalization refers to the evolutionary increase of the relative size of a brain and involves a shift from non-cortical parts of the brain to the cortex. With respect to language processing, this higher encephalization level in humans is probably not a coincidence. It is to assume that a certain neural complexity may be required for the functional complexity involved in human language processing. However, a precise correlation between the child's capacity to acquire effortless and intuitively language and brain size has not been established.

In general, neurodevelopmental processes are either hardwired or activity-dependent. Hardwired processes are predetermined by a genetic program and can be considered as the blueprint for brain's structures. They take place independent of neural activity and sensory data and are expressed at the neuronal level for cell differentiation, cell migration, and axon guidance. After the axons reached the predetermined cortical target area, activity-dependent processes will follow and neural activity and sensory data enable the formation of new synapses and synaptic plasticity. The synaptic plasticity involves the amount of neurotransmitters released into a synapse and represents subject to Hebb's cell-assembly theory, the neurochemical base for learning and memory.[1] According to the cell-assembly theory, associative learning is the product of simultaneous cell activation which leads to synaptic strength between those cells. The brain forms neural networks consisting of neurons, which communicate to each other by using electrochemical signals.

[1] Donald Olding Hebb (1904–1985), a Canadian neuropsychologist, is considered as a pioneer of neural network models. The quote "Neurons that fire together wire together" is known as Hebb's Law. Neurons, which fire together, are considered to be one group or processing unit, called "cell-assemblies."

D. Hillert, *The Nature of Language,* DOI 10.1007/978-1-4939-0609-3_5,

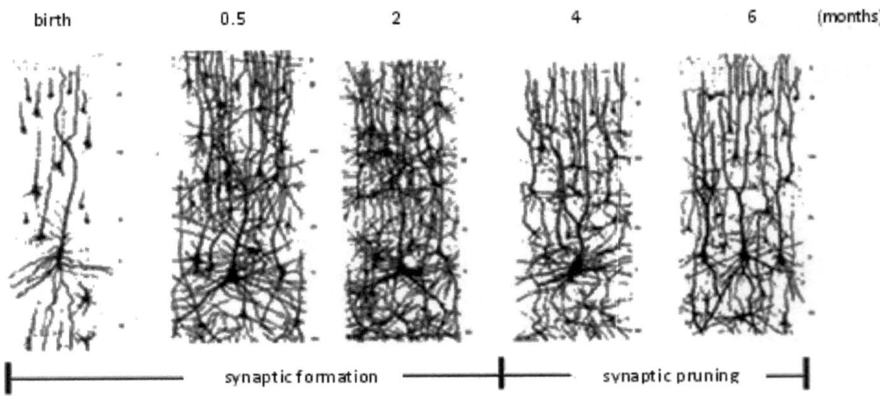

Fig. 5.1 Synaptic proliferation and pruning in children. (Adapted and modified, Seeman 1999)

A neuron has a cell body, from which branch-like structures extend. These structures include dendrites and axons with numerous terminals. The cell body entails the DNA and generates energy, the dendrites receive signals from other neurons and the axons and its terminals carry outgoing signals to other neurons. The signals an axon terminal of one neuron carries to the dendrite of another neuron typically takes place in form of an action potential at a synaptic cleft occurring in less than 1 ms. The action potential is an all-or-nothing process that activates at the presynaptic axon terminal, the synaptic transmission process. In short, an increase of intracellular calcium concentration causes vesicles to release neurotransmitter absorbed by the receptors of the target neuron. The target neuron will be then excited, inhibited, or metabolic changes will occur depending on the type of receptors activated.

Starting in the embryonic stage, new neurons and synapses are formed until approximately 2 years of age at a rate that reaches sometimes 40,000 new synapses per second. Young children are left with more neurons and synapses than cognitively required. Synaptic pruning is an epigenetic factor, which helps to regulate this brain development by the reduction of the total number of neurons and synapses to form efficient cortical circuits. Synaptic pruning starts at birth, a stage at which the human brain consists of more than 10 billion neurons (Craik and Bialystok 2006) but continuous until the age of ca. 10 years (see also Fig. 5.1). Then, almost 50 % of the synapses available at the age of 2 years will be eliminated. The brain size increases at the age of 2–5 years, and about 90 % of the adult brain size has been reached by the age of 6 years (Dekaban 1978), myelination and synaptic pruning are specifically active during this plateau-phase of development, but also during adolescence and early adulthood. A typical synaptic pruning seems to be characteristic for developmental delays of language and other cognitive functions but may be related to a variety of different causes. In typically developing young children, language comprehension relies on a bilateral distributed cortical network. Synaptic pruning seems to stabilize the language network and other cognitive functions resulting gradually in a left-hemispheric lateralization in childhood. The process of synaptic pruning may vary by brain regions, but this line of research requires further longitudinal and/or cross-sectional studies to consider factors related to the maturation of the child's brain. In considering the difference between hard-wired genetic parameters and activity-dependent synaptic pruning, the role of the BDL can be tentatively summarized as follows:

The BDL in a narrow sense may refer to activity-independent processes, while the BDL in a broader sense may involve activity-dependent processes refining neural circuits of language processing. Thus, the BDL, in a narrow and broad sense, can be regarded as the language-genotype. The neurodevelopment expression of the BDL is fixed for all humans, only genetic disorders or atypical sensory deprivations prevent typical language acquisition and development. At the same time, we are aware that biological systems have dynamic properties. The BDL can be considered as a fixed architecture and alterations of this genetic disposition implies a change of what we call human nature. Moreover, in context of the systematic brain changes during a human's life cycle, some systematic linguistic changes can be observed relying on cognitive factors such as working memory or learning capacity. These cognitive alterations modify linguistic performance but do not affect the BDL, neither in a narrow or a broad sense. Also, the activity-dependent process of parameter setting of the child's native language is not a process that prevents the acquisition of other languages at any time during the life cycle. We discuss in a separate chapter whether the acquisition of one or more languages affects the way people processes their native language.

Thereby, the question arises whether the BDL is differently expressed among individuals, that is, whether some individuals are more predisposed for language or specific linguistic tasks than others. Often these possible differences were addressed in context of gender-specific cognitive abilities along with particular brain structures and/or hormone levels. The trend in neuroscience is to emphasize that many aspects linked to gender differences may relate to different factors. This is a fundamental methodological problem in science as cultural bias can distort the attempt to consider alternative interpretations (Jordan-Young 2010; Fine 2010). Male and female brains and behaviors certainly differ in many aspects, but any two experimental groups may have remarkable differences and this applies of course to individuals as well. This phenomenon can be exemplified in the case of a fetus, who is exposed to higher levels of androgens (congenital adrenal hyperplasia—CAH) caused by an inherited enzymatic defect preventing glucocorticoid production. Androgens stimulate and regulate the development of male characteristics in vertebrates. Girls with CAH seem to be better at mentally rotating shapes compared to the unaffected girls. It has been also argued that women have a thicker corpus callosum than men, which would support the observation of greater verbal skills in women than in men. But here, the factor brain size has not been controlled, that is, on average it is the case that the size of the brain correlates with the size of the corpus callosum. The final example refers to the often quoted toy preferences, but who defines what a female or a male toy is? There are many genetic and social factors (e.g., rejection of opposite gender behavior, activities associated with a toy, social shaping) that can trigger a behavioral preference or avoidance in young children. The hormone level can play one important role in behavioral preferences but these changes are gender-independent and need to be considered at an individual level. Thus, the factor "gender" does not play a role in characterizing the BDL. Women and men are born with the same biological capacity for language. Possible individual differences of language abilities may be due to social and educational experiences.

5.2 Linguistic Wiring

Let us focus on the neural architecture of language processing in adults. In context of language-related homologs, we already explored important neural components of the human language system. Today's evidence shows that an extensive network is involved in language processing (e.g., Price 2010; Turken and Dronkers 2011; Friederici and Gierhan 2013). Typically, sentence processing is recruited by a broad left-lateralized network involving the perisylvian cortex and its neighboring regions such as the posterior MTG, inferior temporal regions, inferior partial lobe, IFG, and other frontal regions, which support aWM functions and rehearsal operations. However, most cortical regions mentioned are multifunctional and do not support only language processing. More important is, how these regions share their resources, that is, how they coordinate their activities to provide the functional integration of different cortical regions. The specific involvement of a cortical region in language processing is determined by its connectivity patterns, that is, by the kind of interaction with other regions. These cortical regions can involve not only the left-sided network, but also right-sided regions, if the computational costs are higher for particular linguistic aspects. A precise model of the neural architecture of the human language system is significant for understanding spontaneous neural recovery processes caused by brain injuries or diseases.

To begin with, the typical canonical language stream involves the left IFG and the left superior temporal gyrus (STG). While the production of words and speech recruits motor and premotor regions, the perception of words and speech recruits auditory and occipital regions. At this point, we are drawing a general picture about the human language system, but do not consider specific aspects of language processing. Two ventral routes seem to be in particular relevant: the uncinate fasciculus (UF) and the extreme capsule fiber system (ECFS). UF connects subcortical structures of the anterior temporal lobe (hippocampus and amygdala) with the anterior IFG. The ECFS is a long fiber pathway, which connects the IFG with the middle-posterior regions of the STG and extends to occipital regions. Different kind of evidence including MRI and lesion studies indicate that the ECFS is particularly involved in semantic processing (e.g., Thompson-Schill et al. 1997; Vigneau 2006; Rolheiser et al. 2011). The function of the UF has been attributed to basic or local syntactic computations such as phrase structures. In contrast, while the arcuate fascilicus (AF) seems to be in charge of more complex syntactic processing, the longitudinal fasciculus (SLF) appears to be primarily involved in functions

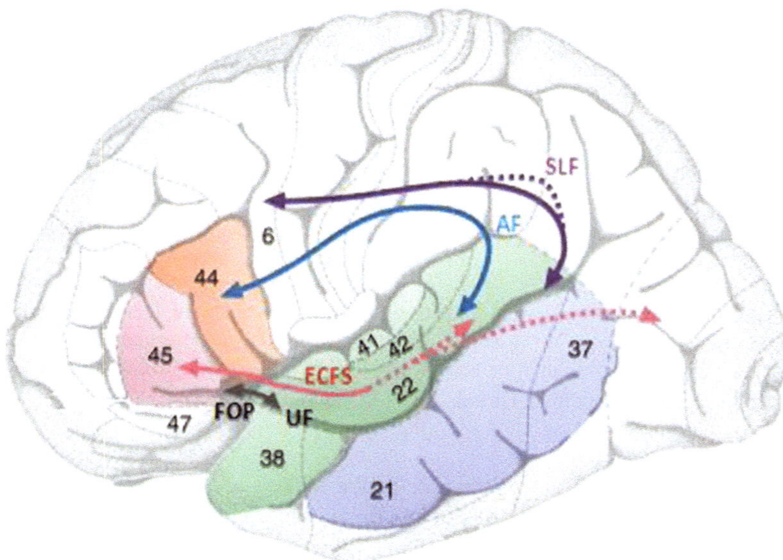

Fig. 5.2 The canonical left-lateralized human language net recruits particular regions by means of dorsal and ventral fiber tracts between posterior and anterior regions. (FOP frontal operculum, UF unicate fasciculus, ECFS extreme capsule fiber system, AF arcuate fascilicus, SLF longitudinal fasciculus). (Adapted and modified, Friederici 2013; © Elsevier Limited)

associated with speech repetition such as perception and production of speech and phonological WM operations (cf. Fig. 5.2).

In adopting evidence from the macaque's brain, the subdivisions SLF I, II, III has been also proposed for the human fiber tract system (e.g., Thiebaut de Schotten et al. 2012). However, their precise linguistic functions still need to be determined.

As mentioned previously, the dual stream model for speech considers temporal factors to account for the combination of different kind of linguistic computations (Hickok and Poeppel 2004, 2007; Hickok 2012a, b). With reference to evidence that indicates a bilateral representation of the ventral stream, it has been speculated that depending on the sampling rate the left (25–50 Hz) and/or the right ventral pathway (4–8 Hz) will be used during word recognition. The assumption is based on the fact that rapid spectral changes (e.g., formant transitions associated with place-of-articulation information) occur in a time window of 20–40 ms, while syllabic and prosodic computations take place in the range of 100–200 ms. Convincing evidence for a bilateral ventral stream for speech, which involves the MTG and STG, stems from lesion studies indicating that speech perception is not significantly impaired (e.g., Rogalsky et al. 2008). In contrast, severe disorders of speech perception results from bilateral lesions of the STG (Poeppel 2001). Again, the proposed left-dominant dorsal stream involves the posterior planum temporale (PT), the premotor cortex and the IFG. The PT[2] is located posterior to the auditory cortex (Heschl's

[2] Comparable to Broca's area, the left PT is typically larger than the one in the right hemisphere and this asymmetry has been also reported for chimpanzees (Carroll 2003). However, left-sided

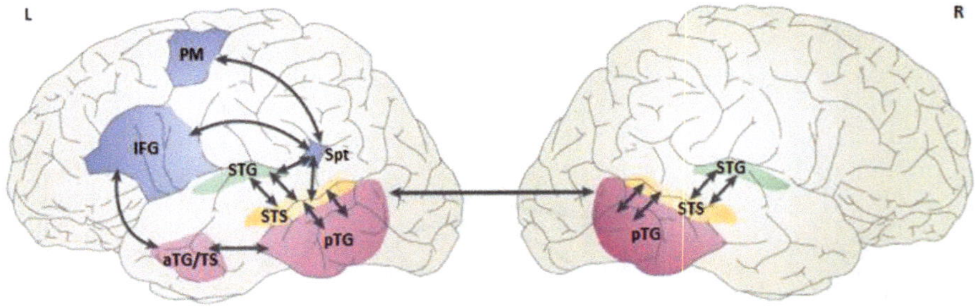

Fig. 5.3 Dual stream model of speech processing. (p posterior, a anterior, d dorsal, mp mid-post, ITS inferior temporal sulcus, MTG middle temporal gyrus, STS superior temporal sulcus, STG superior temporal gyrus, IFG inferior frontal gyrus, PM premotor cortex, Spt Sylvian parietal-temporal). (Adapted and modified, Hickok 2007; © Nature Publishing Group)

gyrus) and can be considered as the core of Wernicke's area as part of the STG and the parietal lobe. In general, the PT seems to be involved in early auditory processing and in absolute pitch recognition[3]. The PT has been also considered as a computational hub, an engine that analyses many types of complex sounds and segregates and matches spectrotemporal patterns (Griffiths and Warren 2002).

The dual stream model makes mainly assumptions about lexical processing in healthy adults but represents also a framework for the prediction of linguistic disorders due to lesions in specific cortical areas. In particular, it is assumed that meanings (lexical conceptual information) are widely distributed throughout the cortex (Hillert and Buračas 2009). In contrast, phonological computations operate in the superior temporal sulcus (STS)[4]. Again, the interface between phonological and conceptual information appears to be primarily the domain of the MTG and STG while semantic-syntactic information at the sentence-level seems to be recruited by anterior temporal lobe areas (e.g., Binder et al. 1997; Friederici et al. 2000; Vandenberghe et al. 2002). The predicted role of the anterior temporal lobe in sentence processing seems to be consistent with Friederici's language network model (Fig. 5.3). Traditionally, linguists differentiate between semantic and conceptual information (in contrast to psychological views) to separate between language-specific (semantic) and general cognitive meaning representations (conceptual). This division seems to be also empirically relevant.

Moreover, the dorsal stream is said to be responsible for auditory-motor integration. In particular, it is believed that in addition to motor areas the regions Sylvian parietal-temporal (Spt) and STS represent the sensory-motor integration system. Hickok (2012) discusses in context of the dual stream model two further assump-

asymmetry of the PT seems to be associated with language laterality (Foundas et al. 1994).

[3] Absolute or perfect pitch refers to a person's skill to reproduce or identify a tone (usually in the domain of music) without hunting for the correct pitch. It appears that the absolute pitch is more common among speakers of tonal languages such as Chinese, Nilo-Saharan, Punjabi, Tai, or Vietnamese.

[4] The STS separates the STG from the MTG.

tions about forward prediction in speech perception. While forward prediction by the ventral stream facilitates speech recognition by generating top-town expectations, possibly mediated by priming and/or context, the forward prediction by the dorsal stream seems to be based on motor functions to enhance speech perception. However, the use of motor-based predictions is obviously not critical for speech perception as damage to the motor speech system does not cause corresponding deficits in speech perception.

We discuss more specific language-related cortical circuits in Part III. In particular, we need to consider that the contribution of the right-hemispheric streams and regions and the limbic system has an important role in sentence processing. Also, the interface of the conceptual system, which appears to be widely distributed, along with components of the language system is essential for forwarding models not only at the phonological but also at syntactic-semantic level. The ability of language processing is based on the BDL. Its implementation requires a certain set of neurobiological parameters, healthy psychological, and mental states and a proper socio-communicative environment. However, the discussion of an atypical language development illustrates that these limits can be quite dynamic and flexible. Although our brain will be functionally structured according to typical patterns, preprogrammed by the BDL, sufficient evidence has been collected indicating that the neural structures supporting language processing are not language-specific. In other words, other cortical regions can take over regions that are typically preprogrammed by the BDL, in particular during the developmental stage. Thus, what seems to be language-specific is that certain linguistic functions are predominately computed by specific cortical areas, a genetically-driven process that replicates to some extent the evolving linguistic and communicative abilities of our biological ancestors.

References

Binder, J.R., Frost, J.A., Hammeke, T.A., Cox, R.W., Rao, S.M., & Prieto, T. (1997). Human brain language areas identified by functional magnetic resonance imaging. *The Journal of Neuroscience, 17*(1), 353–362.

Carroll, S.B. (2003). Genetics and the making of Homo sapiens. *Nature, 422*(6934), 849–857.

Craik, F., & Bialystok, E. (2006). Cognition through the lifespan: mechanisms of change. *Trends in Cognitive Sciences, 10*(3), 131–138.

Dekaban, A.S. (1978). Changes in brain weights during the span of human life: relation of brain weights to body heights and body weights. *Annuals of Neurology, 4*(4), 345–356.

Fine, C. (2010). *Delusions of gender: The real science behind sex differences*. London: Icon Books.

Foundas, A. L., Leonard, C. M., Gilmore, R., Fennell, E., & Heilman, K. M. (1994). Planum temporale asymmetry and language dominance. *Neuropsychologia, 32*(10), 1225–1231.

Friederici, A.D., Meyer, M., & von Cramon, D.Y. (2000). Auditory language comprehension: an event-related fMRI study on the processing of syntactic and lexical information. *Brain and Language, 74*(2), 289–300.

Friederici, A.D., & Gierhan, S.M.E. (2013). The language network. *Current Opinion in Neurobiology, 23*(2), 250–254.

Griffiths, T. D., & Warren, J. D. (2002). The planum temporale as a computational hub. *Trends in Neurosciences, 25*(7), 348–353.

Hickok, G., & Poeppel, D. (2004). Dorsal and ventral streams: a framework for understanding aspects of the functional anatomy of language. *Cognition, 92*(1-2), 67–99.

Hickok, G., & Poeppel, D. (2007). The cortical organization of speech processing. *Nature Reviews. Neuroscience, 8*(5), 393–402.

Hickok, G. (2012a). Computational neuroanatomy of speech production. *Nature Reviews. Neuroscience, 13*(2), 135–45.

Hickok, G. (2012b). The cortical organization of speech processing: Feedback control and predictive coding the context of a dual-stream model. *Journal of Communication Disorders, 45*(6), 393–402.

Hillert, D., & Buračas, G. (2009). The neural substrates of spoken idiom comprehension. *Language and Cognitive Processes, 24*(9), 1370–1391.

Jordan-Young, R.M. (2010). *Brain Storm: The flaws in the science of sex differences*. Boston: Harvard University Press.

Poeppel, D. (2001). Pure word deafness and the bilateral processing of the speech code. *Cognitive Science, 21*(5), 679–693.

Price, C.J. (2010). The anatomy of language: a review of 100 fMRI studies published in 2009. *Annals of the New York Academy of Sciences, 1191*(1), 62–88.

Rogalsky, C., Matchin, W., & Hickok, G. (2008). Broca's area, sentence comprehension, and working memory: an fMRI Study. *Frontiers in human neuroscience, 2*(14). doi: 10.3389/neuro.09.014.2008.

Rolheiser, T., Stamatakis, E.A., & Tyler, L.K. (2011). Dynamic processing in the human language system: synergy between the arcuate fascicle and extreme capsule. *The Journal of Neuroscience, 31*(47), 16949–57.

Seeman, P. (1999). Brain development, X pruning during development. *American Journal of Psychiatry, 156*, 168.

Thiebaut de Schotten, M., Dell'Acqua, F., Valabregue, R., & Catani, M. (2012). Monkey to human comparative anatomy of the frontal lobe association tracts. *Cortex, 48*(1), 82–96.

Thompson-Schill, S. L., D'Esposito, M., Aguirre, G. K., & Farah, M. J. (1997). Role of left inferior prefrontal cortex in retrieval of semantic knowledge: a reevaluation. *Proceedings of the National Academy of Sciences of the United States of America, 94*(26), 14792–14797.

Turken, A. U., & Dronkers, N. F. (2011). The neural architecture of the language comprehension network: converging evidence from lesion and connectivity analyses. *Frontiers in Systems Neuroscience, 5*(1).

Vandenberghe, R., Nobre, A.C., & Price, C.J. (2002). The response of left temporal cortex to sentences. *Journal of Cognitive Neuroscience, 14*(4), 550–560.

Vigneau, M., Beaucousin, V., Hervé, P. Y., Duffau, H., Crivello, F., Houdé, O., Mazoyer, B., & Tzourio-Mazoyer, N. (2006). Meta-analyzing left hemisphere language areas: phonology, semantics, and sentence processing. *NeuroImage, 30*(4), 1414–32.

Chapter 6
Semantics and Syntax

6.1 Sentence Structures

The meaning of sentences represents the interface to syntax, phonology, discourse, conceptual representations/thoughts and is therefore the most controversially discussed level of mental representation in the relevant disciplines such as linguistics, philosophy, psychology, and neuroscience. There is, I believe, still a fundamental confusion or controversy about what the meaning of meaning is. It is a biological axiom that the job of any neural network, which operates as a cognitive system, must be able to represent somehow the "external" world. Our beliefs or thoughts about the world are determined how we are able to represent the world and not how the world is. In opposite to the semantic codes of a natural language, which are full of ambiguities, the innate language of thought or mentalese is supposed to be ambiguity-free (Fodor 1975, 1981, 1987; Pinker 1994). This contrasts with the idea that a single language such as Swedish represents the language of thought (Whorf 1956). The debate about the relationship between language and thought (or mental representations in general) is fundamental for understanding how the human language system operates and is therefore a closely related intertwined topic. In the present context, however, we focus on the meanings encoded in sentences of natural languages, assuming this occurs, and discuss the validity of these approaches in context of neurobiological findings so far available.

Here, the term natural semantics refers to the level of mental representation that maps meanings onto syntactic and/or phonological representations. However, the goal of a wide range of (natural) conceptual and cognitive semantic and formal semantic approaches is to find combinatory rules specific to semantic units (e.g., Jackendoff 1983; Pinker 1989). To begin with, a classical natural semantic view describes the semantic interface in terms of verb arguments, which carry semantic or thematic roles (θ-roles). By definition, every argument is assigned one unique θ-role by the verb (theta criterion), although there may be some exceptions depending on language-specific structures, language model or linguistic theory. Some canonical θ-roles are Agent, Patient, Theme, Goal, Recipient, Experiencer, or Instrument. Although one might believe that the θ-role Agent maps onto the grammatical subject (Subject) and Patient or Theme onto a grammatical object

D. Hillert, *The Nature of Language*, DOI 10.1007/978-1-4939-0609-3_6,

(Object), often this is not the case. For instance, in *The window broke*, the Subject *window* corresponds to Patient (affected entity), and in *Mike liked the picture* the Subject *Mike* has the θ-role Experiencer.

The concept of verb argument structures is controversially discussed in numerous linguistic paradigms. We are addressing here a few implications associated with these different accounts, but the main purpose is to introduce the general approach and to discuss in turn the neural substrates associated with the mental computations of semantic structures. Linguistic theories vary essentially in their degree to which verb argument structures are used to map lexical semantic and/or conceptual information to syntactic information. The core building blocks of a traditional linguistic model are syntax, semantics, and conceptual representations, but the relations between these different levels vary largely. For example, one group of linguists believe that the mental or semantic lexicon determines in form of subcategorization frames, which categorical class a verb (or any other lexical item) can carry into a language-specific grammar (e.g., Baker 1979). The following English examples illustrate that different verb meanings come along with the same or different subcategorization frames. Since most approaches consider the grammatical subject as an external argument, which the verb does not subcategorize, the Subject is not mentioned in the following examples. It is possible to say *X broke* or *X broke Y*; the first sentence does not express an object noun phrase (NP), but the second one does with the variable Y. Again, the frames for the sentences *X hit the door* and *X hit the freeway* are identical as both are subcategorizing an object NP, but expressing different kind of meanings, a literal and a figurative one; or let us look at sentences such as *X kicked the door* and *X kicked the bucket*: While in the former sentence only an object NP subcategorization is meaningful, the latter one is ambiguous and permits both frames [_NP] and [_]. The idiomatic reading of *kick the bucket* does not categorize an object NP, but can be considered as a different verb entry. Again, the verb *give* obligatorily subcategorizes two elements, accusative and dative case, but allows two different frames; *X gave Z to Y* [_ NP PP] or *X gave Y Z* [_ NP NP]. In the last example, four different frames are shown for the verb *love*; an NP in *X loves Y*, a subclause with an infinitive verb (_ S'_ INF) *X loved to write*, an S' with an *ing*-verb *X loved writing* [_ S'_ING] and in addition with an object NP *X loved him writing* [_ NP S'_ING]. Other approaches postulate that subcategorization frames are redundant as the verb argument structure can be derived from the verb's meaning (e.g., Pinker 1989; Levin 1993). The construction account, however, emphasizes that the verb's meaning alone does not inform about the meaning of a sentence (Fillmore et al. 1988; Goldberg 1995), for instance, sentences such as *They laughed the poor guy out of the room* or *Frank sneezed the tissue of the table* (Goldberg 1995). The verbs *laugh* and *sneeze* do not encode on their own the meaning of Caused-Motions. It is said that the verbs' core meanings are fused with the argument structure construction, here the Caused-Motion construction.

Here, the view is taken that (natural) semantics is independent of syntax much like phonology or even vision (Jackendoff 2007). Interface rules must maintain the correspondence between these different types of mental representations. In the brief dialogue *Let's travel to Hawaii!* (A) *How about North Shores?* (B), speaker

B mutually agrees on semantic-pragmatic grounds with speaker *A* by suggesting a more precise location. In contrast, in the dialogue *Angélique travelled to Hawaii* (A) *Yes, to Hilo!* (B), both sentences are linked by an ellipsis, that is, the underlying syntactic structure (also called deep structure or logical form) of the B's statement is *Angélique travelled to Waikiki* is deleted and marked as non-pronounced (Merchant 2001). Since such conceptual information encoded in sentences has its own syntax, it may exist without linguistic representations, also in non-human primates, who are able to express intentions without being able to speak. The boundary conditions for a model of conceptual semantics are hypothesized as follows (Jackendoff 2007; see also Jackendoff and Pinker 2005): (a) sentences must be semantically composed by integrating lexical syntactic information; (b) sentential semantics serves as basis for inferences; (c) lexical meanings must conform to conceptual categorizations; (d) lexical meanings must be learnable based on experience with language and the world; (e) sentence meanings expressing physical objects or actions must correspond to mental representations of perception or actions; (f) pre-linguistic thoughts or mental structures underlying sentence meanings can be found to some degree in nonhuman animals and play a role in understanding the world.

It is acknowledged that the paradigm of conceptual semantics is similar to cognitive grammar (e.g., Langacker 1987; Fauconnier 1985; Fauconnier and Turner 2002) as the cognitive-psychological nature of meaning is enquired. While the approach of conceptual semantics commits to syntax and is designed to be independent of linguistic semantics, the approach of cognitive grammar tries to describe all syntactic aspects in terms of meanings. Thus, the goal of formal semantics is to describe meanings in terms of formal-logic axioms, but not to model semantic structures encoded in natural languages (Hillert 1987, 1992). This implies that formal semantics is in general not useful for the investigation of the mental reality of natural semantics. In order to avoid being dogmatic, however, in principle it cannot and should not be excluded that some formal inferences on specific semantic aspects might be relevant or valuable for natural linguistic approaches when developing contrastive hypotheses (even if they are negative examples). In this vein, computational approaches such as latent semantic analysis (LSA) (Landauer 2007) turned out to be quite valuable for determining the probability of a word co-occurring with other words in a document, but no relationship is expressed to the underlying mental operations responsible for the generation of semantic representations. In contrast, Wierzbicka's (1996) compositional approach postulates universal semantic primitives such as mental predicates (*think, know, want*) or (pro-)nouns (*I, you, person*), which are supposed to represent the innate structure for the acquisition of lexical meanings. A set of rules combines these prime concepts and represents a prime language for natural semantics (meta-language) (Fig. 6.1).

However, it remains unclear and it would need to be spelled out how this semantic meta-language can be mapped onto the semantics and syntax of sentences. Other related approaches developed in the domain of psychology tend to frame the structure of lexical concepts as part of knowledge representations including episodes, schemas, scripts, features, prototypes, and basic concepts rather than of sentential semantics. A semantic approach should not only consider the interface to syntax

Fig. 6.1 Semantics and related representations

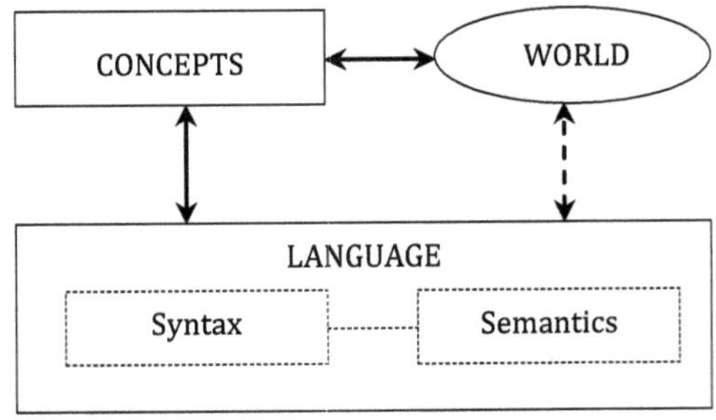

and conceptual representations but also the interface(s) of the conceptual system to the world and possibly a direct connection between sensory data and the language system.

At the level of syntactic representations, sentences (and possibly dependencies across different sentences) are decomposed into different phrases, including intermediate phrases, according to general principles. These principles also imply syntactic recursion and they can be therefore illustrated in form of tree structures. For instance, the rule S -> NP VP informs that a sentence can be generated by an NP followed by a verb phrase (VP). The X-bar (X') theory developed by Chomsky (1970) and Jackendoff (1977) is used in many grammatical models and describes syntactic similarities across all natural languages. X is a placeholder for syntactic categories and dependent on the level of the syntactic hierarchy, X^0 refers to the head (e.g., N, noun), to intermediate levels (X') or to a phrase (X'', X-bar-bar). All phrases are projected from lexical categories in the same way. Typically, phrases are diagrammed as tree structures and branching is always binary. Let us looks at a basic structure: The maximal projection X'' branches into the intermediate projection X' (daughter) and a specifier (Spec) and X' branches into X^0 and a complement (Comp). Multiple sisters of X' can be generated (recursion) such as adjuncts (Adj).

(3)

a. b.

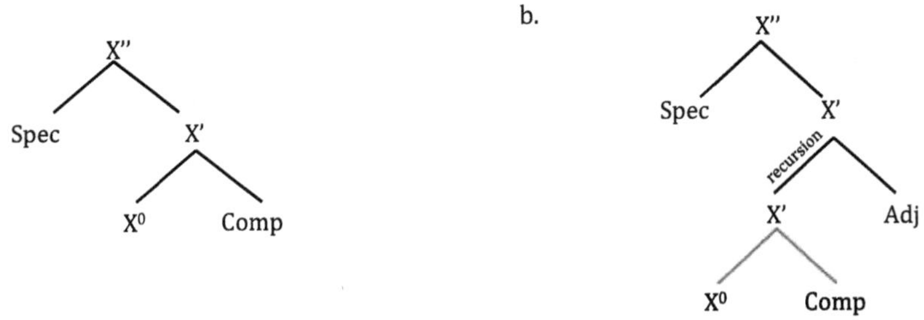

Thus, structure (3a.) can generate phrases such as the *author* (Spec X) or *always writes books* (Spec X Comp) or structure (3b.) generates the phrase *always writes books in the cabin* (Spec X Comp Adjunct). Based on the X' schema other theories followed, government binding (GB) and minimalism, to account for relatively complex syntactic clauses but also for referential expressions, tense or Case

(Chomsky 1981, 1995, 2005). As the underlying syntactic structures are considered to be universal, the schema applies to word orders of any natural language. For example, in considering X" -> Spec X', X' -> X^0 Comp applies to an SVO (subject verb object) language (e.g., English, Portuguese, Russian, partly Chinese) and X' -> Comp X^0 to the SOV order (e.g., Japanese; Hindi, Urdu, partly German); in contrast, in considering X" -> X' Spec, X' -> X^0 Comp applies to the VOS order (most Austronesian languages such as Tagalog or Fijian) and X' -> Comp X^0 to OVS (mostly due to case markings such as German, Finish, Romanian, Hungarian).[1] Moreover, mapping D-structures onto S-structures will be shown with co-indexed traces *(t)* indicating the movement of lexical category out of its canonical position in a sentence. Below in (4) the S-structure of the question *Which book has the author written?* is illustrated, whereas C" is a complement phrase and I" the inflection phrase.[2]

(4)

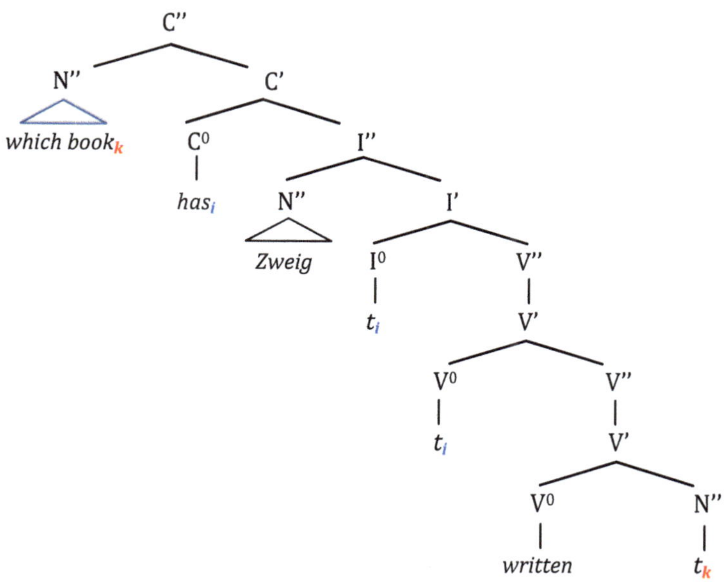

In the GB account, θ-roles, which are part of subcategorization frame, cannot move out of the original D-structure position. This principle can be illustrated with passive sentences. For instance, in the sentence *Dita was invited to the opera by Angélique*, the lexical entry of the verb includes the thematic grid <Theme, Goal, Agent> and

---

[1] The word orders VSO and OSV also occur, whereas OSV (e.g., Urubú, Brazil) is relatively rare (see Chung 1990; McCloskey 1991) for a discussion of these structures in context of the X' schema.

[2] For simplification, we do not include here grammatical features such as agreement, tense or type of NP.

the subcategorization frame [_ NP ($PP_{[to]}$) ($PP_{[by]}$)]. Mapping D- to S-structure of the passive sentences involves movement of the Object (Dita) into the Subject position. Since the Subject has no θ-role in the D-structure, no information gets lost. Using the bracket notations, the D-structure and S-structure of the passive sentence mentioned above are respectively shown in (5a/b) and in addition (6) diagrams the S-structure. Here, the arguments are marked by letters, whereas the daughters of P'' and N'' are not spelled out (A, Angélique; B, Dita; C, the opera):

(5) a. $[_{I''}[_{I'}[_{I^{o}[+past]}[_{V''}[_{V'}[_{V^{o}}\ \textit{was}][_{V''[+past]}[_{V'}[_{V^{o}[+pass]}\ \textit{invited}][_{N''}\ \textit{B}][_{P''[to]}\ \textit{to C}][_{[P'']}\ \textit{by A}]]]]]]]$

b. $[_{I''}[_{N''}\ \textit{B}_{k}][_{I'}[_{I^{o}}\ \textit{was}_{i}][_{V''}[_{V'}\ [_{V^{o}}\ t_{i}][_{V''[+pass]}[_{V'}[_{V^{o}[+pass]}\ \textit{invited}\ ][_{NP}\ t_{k}][_{P''[to]}\ \textit{to C}][_{[P'']}\ \textit{by A}]]]]]]$

(6)

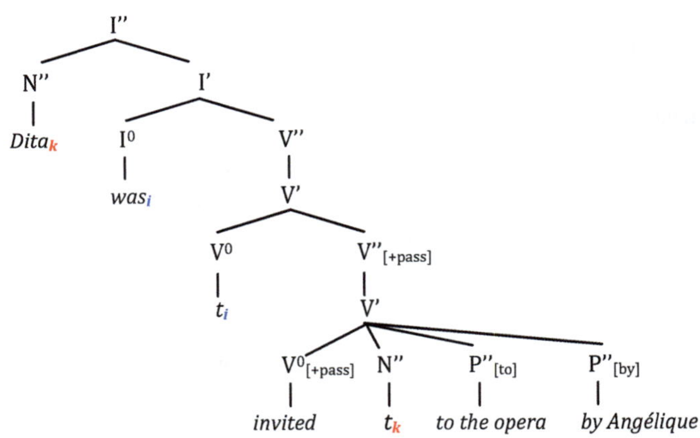

A series of other syntactic rules are expressed in GB such as the binding theory to handle referential expressions or the Case theory that assumes abstract Cases for all N''. Here, we simply sketched the analysis of some syntactic structures in the paradigm of generative grammar to provide the reader with an idea about this mentalistic approach. The analysis of the intrinsic operations at the semantic or syntactic level and their interfaces is a complex enterprise. Alternatively, more recent syntactic theories such as lexical functional grammar (LFG), head-driven phrase structure (HPSG) or dependency grammar (DG) project grammatical relations from lexical rules (e.g., Mel'čuk 1987; Pollard and Sag 1994; Heringer 1996). For instance, the idea of a DG, which goes back to structural linguistics (Tesnière 1959), is that the syntactic structure of a sentence consists of binary asymmetric relations. D-relations can be compared to an X' schema, but which has only one level, that is, the dependents of a word are the heads of its sisters and dependency relations between two words are unified. For illustrative purposes, let us use the passive sentence mentioned above (7a). The dependency relations are shown in (7b) and illustrated in (7c). As DG is verb-centered, the verb is considered as the "root" and not separately mentioned.

(7) a. Dita was invited to the opera by Angélique.

b. nsubjpass (*invited, Dita*); auxpass (*invited, was*); prep_to (*invited, to*); det (*opera, the*); pobj (*to, opera*); prep_by (*invited, by*); pobj (*by, Angélique*) (Abbreviations: nsubjpass (noun subject passive), auxpass (auxiliary passive), prep (preposition), pobj (prepositional object), det (determiner)).

c.

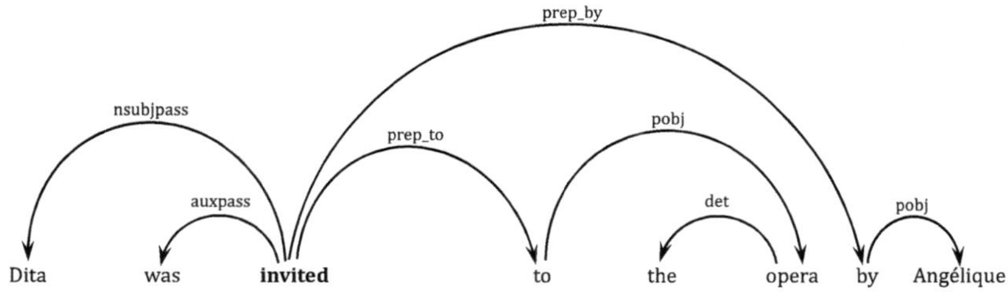

The passive sentence (6a) consists of 8 words and 7 word pair dependencies as shown in (7b, c). The internal linguistic discussion debates the benefits of a particular syntactic theory according to specific (cross-)linguistic structures, but essentially all approaches try to cover the same phenomena. Often it is a matter of the philosophy to what extent semantic or even conceptual information should be considered and whether a theory is sufficiently rich to provide testable hypotheses. It is not uncommon that sentence processing data are compatible with a range of different cognitive–linguistic theories or models. Linguistic theories at various levels of representations are doubtless valuable contributions from a methodological viewpoint. However, ultimately the goal is to develop a language theory that bridges or merges cognitive and neural structures. A sample approach is a dual-domain syntactic theory account that, for example, maps syntactic structures and rules in language and music (e.g., Lerdahl and Jackendoff 1983). To discover common principles of human cognition it may be important that the object of the investigation is not restricted to the language domain. A global theory of human cognition may be required, which is broad enough to cover general parameters of different domains (perhaps including nonhuman cognition), but which is also specific enough to meet domain-specific parameters.

6.2 Neural Nets

The main elements of a neural net(work), which try to simulate neurobiological processes of brain functions, consist of chemically and/or functionally associated neurons. Each single neuron has synaptic axon-to-dendrites connections to many other neurons involving electric and neurochemical signaling. Cognitive models or artificial intelligence (AI) theories are often inspired by neural nets to simulate biological-cognitive behavior or to develop software systems (e.g., Maltarollo et al., 2013). While the direct benefits of an AI account for the simulation of brain functions is debatable, the purpose of cognitive or functional net models is to simulate those cognitive functions according to the structure and function of neural networks (nets). Neural nets are, for example, used in connectionist models, which is appealing as notations and computations are comparable to those found in neurobiologically motivated nets. Another important benefit is that connectionist models simulate cognitive behavior across different domains.

The parallel distributed processing (PDP) account of connectionism is based on the following principles (Rumelhart et al. 1986): (a) mental representations are parallel or distributed activities involving patterns of numerical connections; (b) the acquisition of mental representations results from the interaction of innate learning rules and architectural properties; (c) connection strengths will be modified with experience. It was one of the first attempts to explain cognitive behavior apart of rules and symbolic representations by using the supervised learning algorithm back-propagation of errors (backprop) in a multi-layer perceptron without any loops. It is a feed forward neural net, that is, the data flow only in one direction from input to output.As neural nets are inspired by biological structures, the nodes in an artificial neural net mirror to some extent neurons. The output represents each node's activation or values and these weight values determine the relation between input and output data. Weight values are the result of a training phase, in which data iteratively flow through the network.

Let us briefly look at the biological neuron analogy (Fig. 6.2). The dendrites are input nodes, which receive information from neurons in the previous layer. They transfer the information to the cell body, the soma, and in turn information will be output to the axon, which carries this information to other neurons via synapses. Again, at the synapses the terminals of an axon are connected to the dendrites of neurons in the next layer.

An artificial neuron (node) has, much like a biological neuron, multiple inputs. Soma and axon are replaced by a summation and transfer function and the output serves as input for multiple input nodes.

For instance, when the neuron is fed with input data, the summation function calculates the net value by multiplying the input value (x_i) with the associated connection weights (w_i): net value = $\sum_{i=0}^{n} w(i)\,x(i)$. Again, the transfer function[3] uses the

[3] A transfer function can be for instance sigmoidal or hard limited. The sigmoid function takes a net value and generates an output between 0 and 1 and a hard limited function sets for example a fixed range such as $<0.5=0; \geq 0.5=1$.

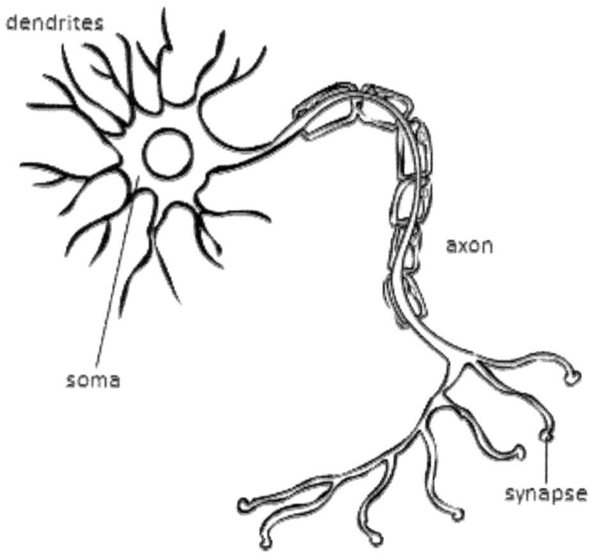

Fig. 6.2 Simplified schema of a biological neuron. (Adapted and modified; @ Maltarollo et al. 2013)

net value to produce an output, which will be then propagated to the input nodes of the next layer. In a backprop network, the error (delta) function is used to calculate the difference between the targeted and actual output. Weights and biases are then adjusted and this iterative process may lead to an improved net output that targeted and actual output match in most of the cases. In contrast to a feed forward architecture, the well-known "Elman network" consists of a multilayer learning algorithm (perceptron) with an additional input layer—the context layer (see Fig. 6.3a–c; Elman 1990; see also Jordan 1986; Hertz et al. 1991). This context layer receives as input a copy of the internal states (hidden layer activations) at the previous time step. Thus, the internal states are fed back at every time step to provide a new input and data about its prior state. In principle, this recursive function or feedback connections can keep a pattern indefinitely alive.[4] The recurrence provides dynamic properties and enables the net to compute sequential input data. The connection weights are random at the beginning of the training session and it has to find over time how to encode the structure internally. Thus, the network does not use any linguistic categories, rules or principles.

Elman (1991, 1993) trained this net with relatively complex sentences, that is, they could have multiple relative clauses and included verbs with different argument structures. In the training session, each word received a vector of zeros (0s), in which a single bit was randomly set to 1. Overall, the net did not succeed, sentences such as *The boy who the girl *chase see the dog* were predicted.

[4] It should be emphasize that connectionist models include distributed (non-modular) but also sequential (modular) computations, in which different types of data are represented over different groups of units allowing the direct interaction of only specific datasets as in simple recurrent networks.

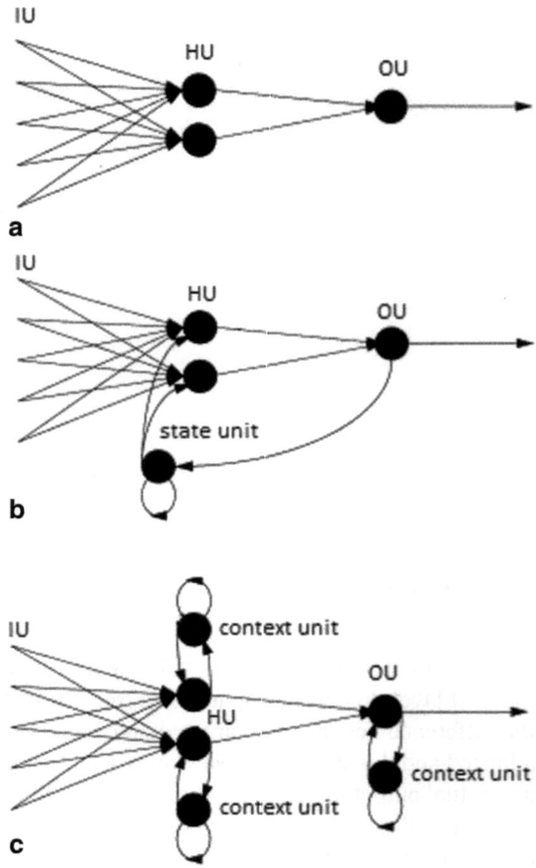

Fig. 6.3 Neural net architecture: **a** multi-layer perceptron **b** Jordan network **c** Elman network (IU input unit; HU hidden unit; OU output unit). (Adapted and modified, Tiira 1999; © Elsevier Limited)

However, when the net was trained first on a restricted datasets in several stages from simple to more complex sentences, the output was quite successful as even long-distance dependencies could be handled. Thus, neural nets can quite well simulate sentence processing as it learns from statistical co-occurrences. Can we therefore say that recurrent nets discover linguistic rules? Yes, we can say this as such as these patterns or regularities are considered as rules. Every time, the net receives a new input, the hypotheses of the rule will be evaluated and updated, it is a dynamic learning process. However, can we say that this stepwise approach simulates language acquisition in children? Certainly not, also emphasized by Elman (1993), as the child is instantly confronted with the full range of different types of sentences (adult language), although adults adjust to some extent their conversation to "child language." Thus, the stimulus input is constantly rich and does not change, but what is changing during acquisition is the child's neural net. The learning process involved in the stepwise implementation of simple-to-complex sentences mirrors the biological sentence acquisition process in a child. Neural net approaches trying to simulate the child's acquisition process should keep the external input constant, but would need to implement hidden computations that continuously adapt to more complex sentence structures. Neural net accounts of language learning and processing are promising approaches to simulate intrinsic computations—particularly if in addition electrophysiological and neuroanatomical factors are considered—to which we would find otherwise no access. It might be not necessarily unfavorable that most connectionist approaches do not insists on neurophysiological adequacy of their accounts. However, some approaches as we will discuss below consider more closely neurobiological properties to mimic language processing according to actual brain activities.

A language theory, which makes claims about the biological properties of the human brain, should try to simulate human brain processes. One approach is to map as close as possible linguistic or cognitive behavior to neurophysiological components and operations. Let us review briefly: The basic structure of the human brain consists of a net of neurons, whereas neurons constantly fire at a low rate. Other cells such as glia cells seem to play any role in concept development. Most cortical and some subcortical areas include cell assemblies (CA) or sometimes called neuronal assemblies. Although there is no common agreement on the definition of CA, there is no doubt about their importance for understanding the neurophysiology of cognition. According to Hebb (1949), CA represents the neural representation of a concept. For example, understanding the word *apple* implies that a sufficient number of neurons fire, leading finally to the firing of an organized collection of cells, a cascade of neurons. Thus, the CA of the word *apple* does not include nodes but an associative network. The CAs overlap depending on shared lexical meanings (e.g., *apple–orange*). This persistent firing is associated with working (or short-term) memory functions. The synaptic strength of neuron connections is determined by repeated stimuli leading to the formation of long-term memories (Hebbian learning). It estimated that the CA sizes range between 10^3 and 10^7 neurons, whereas the human brain consists in total of about 10^{11} neurons (Smith 2010). The following principles apply: A relatively small set of CA encode each concept; a single neuron can be a member of different CA. Neurons of a CA showing self-persistent activities, called reverberation. A CA is learned when it consists of a specific set of neurons.[5]

Only a few approaches discuss the idea of modeling sentence processing in terms of CA (e.g., Pulvermüller 1999, 2002; but see Bierwisch 1999). The general idea is that for example grammatically well-formed structures are detected by nets when the sequence of neuronal elements matches the input. The relevant sequence nets would need to detect syntactic categories of words and establish the relations between the words in a sentence. One sequence net would establish the relationship between a noun and a determiner, for example, and the other one the relationship between the noun and a verb. The philosophy behind this approach is to understand language processing in terms of neural components. Many linguistic phenomena, in particular morphological and syntactic patterns, cannot be simulated by an associate

[5] Part of the development of a CA is neural or synaptic death and growth.

network *per se* as represented by an (extended) Hebbian model. Genetic principles specific to a particular species establish neural wiring during cognitive development, a process based on principles different from associative learning *per se*. Independent of the specific neural model, at present the effort should be applauded to map cognitive functions as close as possible to neurobiological structures. A different approach consist in simulating neuronal processes and thus cognitive behavior in an electronic fashion (reverse engineering). For instance, the ultimate goal of the so-called Blue Brain Project is to build a synthetic human brain that reaches the molecular level. Detailed microcircuits of the different brain regions might be simulated to offer a dynamic virtual brain library for different species (e.g., Markram 2006). To what extent these software models are able to simulate complex local and global cognitive processes at the neurochemical level remains to be seen.

References

Baker, J. (1979). Trainable grammars for speech recognition. In D. H. Klatt & J. J. Wolf (Eds.), Speech Communication Papers for the 97th Meeting of the Acoustic Society of America, pp. 547–550.

Bierwisch, M. (1999). Words in the brain are not just labelled concepts. *Behavioral and Brain Sciences, 22*(2), 280–282.

Chomsky, N. (1970). Remarks on norminalization. In R. Jacobs & P. Rosenbaum (Eds.), *Readings in English transformational grammar*. Waltham: Ginn.

Chomsky, N. (1981). *Lectures on government and binding*. Dordrecht: Foris Publications.

Chomsky, N. (1995). *The minimalist program*. Cambridge: MIT Press.

Chomsky, N. (2005). Universals of human nature. *Psychotherapy and Psychosomatics, 74*(5), 263–268.

Chung, S. (1990). VP's and verb movement in Chamorro. *Natural Language and Linguistic Theory, 8*(4), 559–620.

Elman, J. L. (1990). Finding structure in time. *Cognitive Science, 14*(2), 179–211.

Elman, J. L. (1991). Distributed representations, simple recurrent networks, and grammatical structure. *Machine Learning, 7*(2–3), 195–225.

Elman, J. L. (1993). Learning and development in neural networks: The importance of starting small. *Cognition, 48*(1), 71–99.

Fauconnier, G. (1985). *Mental spaces*. Cambridge: MIT Press.

Fauconnier, G., & Turner, M. (2002). *The way we think*. New York: Basic Books.

Fillmore, C., Kay, P., & O' Connor, M. (1988). Regularity and idiomaticity in grammatical constructions: The case of let alone. *Language, 64*(3), 501–538.

Fodor, J. A. (1975). The language of thought. New York: Crowell Press.

Fodor, J. A. (1981). *Representations*. Cambridge: MIT Press.

Fodor, J. A. (1987). *Psychosemantics*. Cambridge: MIT Press.

Goldberg, A. (1995). *Constructions: A construction grammar approach to argument structure*. Chicago: University of Chicago Press.

Hebb, D. O. (1949). *The organization of behavior*. New York: Wiley.

Heringer, H. (1996). *Deutsche Syntax dependentiell*. Tübingen: Stauffenburg [German].

Hertz, J. A., Palmer, R. G., & Krogh, A. S. (1991). *Introduction to the theory of neural computation*. Redwood City: Addison-Wesley.

Hillert, D. (1987). Zur Mentalen Repräsentation von Wortbedeutungen: Neuro- und Psycholinguistische Überlegungen [German]. Tübinger Beiträge Linguistik 290: Tübingen: Gunter Narr Press. (The mental representation of word meanings: neuro- and psycholinguistic considerations)

Hillert, D. (1992). Lexical semantics and aphasia: A state-of-the-art review. *Journal of Neurolinguistics, 7*(1–2), 23–65.

Jackendoff, R. (1977). *X-bar-Syntax: A study of phrase structure*. Cambridge: MIT Press.

Jackendoff, R. (1983). *Semantics and cognition*. Cambridge: MIT Press.

Jackendoff, R. (2007). A parallel architecture perspective on language processing. *Brain Research, 1146*, 2–22.

Jackendoff, R., & Pinker, S. (2005). The nature of the language faculty and its implications for evolution of language (Reply to Fitch, Hauser, and Chomsky). *Cognition, 97*, 211–225.

Jordan, M. I. (1986). *Serial order: A parallel distributed processing approach*. ICS UCSD No. 8604.

Landauer, T. K. (2007). *Handbook of latent semantic analysis*. Mahwah: Lawrence Erlbaum Associates.

Langacker, R. W. (1987). *Foundations of cognitive grammar: Theoretical prerequisites* (Vol. 1). Stanford: Stanford University Press.

Lerdahl, F., & Jackendoff, R. (1983). *A generative theory of tonal music*. Cambridge: MIT Press.

Levin, B. (1993). English verb classes and alternations: *A preliminary investigation*. Chicago University Press.

Maltarollo, V. G., Honório, K. M., & Ferreira da Silva, A. B. (2013). Applications of artificial neural networks in chemical problems. In K. Suzuki (Ed.), *Artificial neural networks—architectures and applications*, InTech, doi:10.5772/51275.

Markram, H. (2006). The Blue Brain Project. *Nature Reviews Neuroscience, 7*, 153–160.

McCloskey, J. (1991). Clause structure, ellipsis and proper government in Irish. *Lingua, 85*(2–3), 259–302.

Mel'čuk, I. A. (1987). *Dependency syntax: Theory and practice*. Albany: State University Press of New York.

Merchant, J. (2001). *The syntax of silence*. Oxford: Oxford University Press.

Pinker, S. (1989). *Learnability and cognition: The acquisition of argument structure*. Cambridge: MIT Press.

Pinker, S. (1994). *The language instinct*. New York: Harper.

Pollard, C., & Sag, I. A. (1994). *Head-driven phrase structure grammar*. Chicago: University of Chicago Press.

Pulvermüller, F. (1999). Words in the brain's language. *The Behavioral and Brain Sciences, 22*(2), 253–279.

Pulvermüller, F. (2002). *The neuroscience of language: On brain circuits of words and serial order*. Cambridge: Cambridge University Press.

Rumelhart, D. E., McClelland, J. L., & the PDP Research Group (1986). *Parallel distributed processing: Explorations in the microstructure of cognition* (Vol. 1). Cambridge: MIT Press.

Smith, K. (2010). Do the billions of non-neuronal cells in the brain send messages of their own? *Nature, 468*, 160–162.

Tesnière, L. (1959). *Éléments de syntaxe structural* [French]. Paris: Klincksieck.

Tiira, T. (1999). Detecting teleseismic events using artificial neural networks. *Computers & Geosciences, 25*(8), 929–938.

Whorf, B. L. (1956). *Language, thought, and reality: Selected writings of Benjamin Lee Whorf.* Eastford: Martino Fine Books.

Wierzbicka, A. (1996). *Semantics: Primes and universals*. Oxford: Oxford University Press.

Chapter 7
Lexical Concepts

7.1 Constructions

One of the most relatively influential nativist account in linguistics and philosophy is Fodor's (1975, 1987, 1998) language of thought paradigm. Fodor argues that lexical concepts are innate and that only complex lexical meanings may be learned. The point is that words are not explicitly learned by experience, but lexical semantic acquisition is triggered by experience. This is considered, as Fodor puts it, a "brutal-causal" rather than a rational process. Even today, the idea is controversially debated among those, who try to understand how meanings of words are represented and/or constructed. Concepts lexicalized or not are subsentential mental representations used to construct thoughts. As the nativist account claims that most concepts are innate, a speaker with a command of let's say 60,000 words would have access to 60,000 innate concepts. To estimate the vocabulary of a person is difficult. First, it needs to be defined what a word is; for example, the entry for the word *set* in the Oxford English Dictionary has itself 60,000 different entries, although they express closely related meanings. Again, if we include for example slang, dialects, proper names, abbreviations, domain-specific, technical terms, or even non-native lexical entries, the size of the mental lexicon is difficult to estimate and may vary largely from person to person. Most of all, the mental lexicon is not a fixed repertoire as people continuously create new lexical entries to describe or refer to new situations or entities.

Many instances, however, indicate that not all lexical concepts are innate. Opponents of this view may argue that scientific terms or any new inventions and creations require learning a concept by definition. A concept may be acquired by constructing it from other concepts, which might be more primitive concepts. However, philosophers failed to provide necessary and sufficient conditions for defining a concept. Even the often quoted *bachelor* example is not a good example for a definitional concept as Lakoff (1987) pointed it out: the meaning of the word *priest* may meet the criterions of the bachelor concept (male, adult, and unmarried) but intuitively a priest is not considered as a bachelor. Thus, semantic definitions appear not to be psychologically real or relevant in language converse (e.g., Van Langacker

D. Hillert, *The Nature of Language*, DOI 10.1007/978-1-4939-0609-3_7,

1987; Croft 1993; Fauconnier 1997; Evans 2004). Our experience initiates concept formations, a capacity, which is species-specific. Definitions may help us in a conscious way to understand our usage of particular words, but these are reflections of our intuitive conceptual capacity. It is a highly controversial and fundamental topic, but the assumption is that words do not have meanings as their meanings are dependent on the context at any given point of time. Word meanings as function of language use are constructed and make use of any kind of experience stored, including common sense and episodic information. Let us look at a few examples: In (8) the verb *catch* expresses three different meanings depending on its context; (8a) refers to a specific action, which depends on the characteristics of the physical entity, in (8b) the actor are the *eyes* (direct object), which *catch* not a physical entity but attention, and (8c) expresses a phrasal expression as nothing will be caught and a person appears at the *wrong time*.

(8) a. He caught the fish.
b. She caught his eyes.
c. He caught her at the wrong time.

Concrete nouns, for example, alter their meanings depending on their modifiers:

(9) a. This is a long train.
b. This is the first train.
c. This is my train.

(9a) refers to the physical dimension of a *train*, (9b) to the schedule of the trains and (9c) typically to the personal schedule of a traveler; finally, let's also provide examples for adjectives:

(10) a. This is a dirty solution.
b. This is a dirty car.
c. This is dirty money

In (10a) the meaning of *dirty* refers to rough or bold, in (10b) to unclean, a perceptual feature of an entity, and in (10c) to illegally acquired entity. Certainly, language makes use of conventional expressions, usually highly frequent, but they can be considered as conversational routines in context of the generation of constructional meanings. A context-induced shift of lexical meanings generates (sub-)sentential meanings, which taps into encyclopedic knowledge of a perceiver. Thus, the interpretation of a sentence is never identical between two perceivers but represents a semantic approximation. Independent of the question of how stored lexical meanings are acquired, lexical representations is the base for situated language use. The key question, thus, is how do lexical representations map conceptual knowledge or structures at the sentence level? Most linguists would agree that only certain aspects of conceptual structures are explicitly encoded in language. Conceptual structures may be regarded as a systematized output of domain-specific sensory modules. They may consist of conceptual fields representing relational and contextual information, conceptual primitives derived from similarities between different domains and inference characteristics. Any new lexical meanings are based on restructuring or blending of existing concepts (Turner and Fauconnier 1995):

(11)

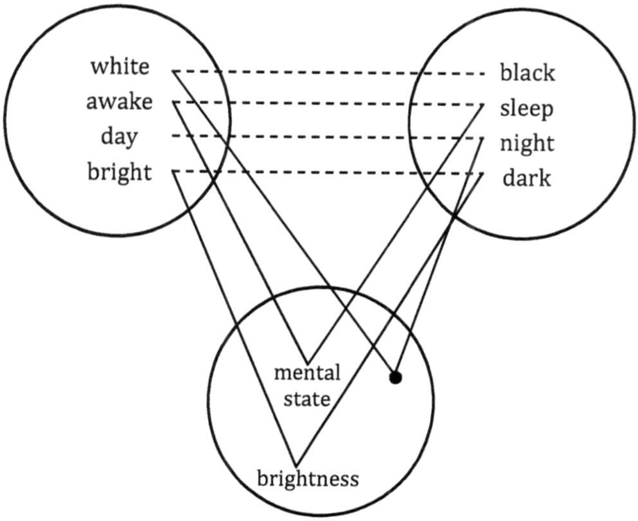

In the case of blends of conceptual fields or spaces, as shown in (11) analogs are mapped that the lexical units do not contradict each other but create new meanings, in this case a figurative meaning. The color term *white* receives a connotative meaning in combination with the head word *night* of the compound *white night. White* refers to awake, active bright that turns the night into a day, although the activities of celebrating may be more suitable for the night than for a day, when we exclude the stereotypical behavior of sleeping at night. Again, this example and many other cases of our daily linguistic experience illustrate that new lexical meanings are created on the go on the basis of a highly dynamic contextually-driven conceptual system.

Rationalists of the seventeenth century such as René Descartes, Gottfried Wilhelm Leibniz or Antoine Arnauld paid particular attention to the need of semantic primitives grounded in language. In modern times, generative linguists such as Jerrold Katz, James McCawley or Ray Jackendoff used this paradigm to identify semantically basic words across all human languages, which cannot be defined further and are thus considered as primitives. But when do we know that the bottom has been reached? Methodologically, does the concept of primitives imply an infinite regress? Let's look at a nonlinguistic example: Atoms were once considered as the smallest building blocks of matter, but further research decomposed them into a nucleus and electrons, the nucleus was further divided into neutrons and protons and both then were decomposed into quarks while quarks are today considered as a set of features (spin, color, charm, etc.).

The search for smaller but universal elements seems to be an essentially cognitive operation of human thinking. Further specification and differentiation provide theories more fine-grained and gain is out of question, at least in the domain of physics. The idea of developing a theory of semantic primitives is to describe more precisely the interface between lexical concepts and sentence information. Generative or cognitive linguists assume that universal or a finite set of semantic primitives are used for the generation of propositions, that is, of sentence meanings. But how to understand these mechanisms more precisely, is the task of future work. There is, however, sufficient evidence that those semantic primitives cannot consist of a fixed set of concepts as novel meanings and concepts are continuously created. Thus, semantic primitives may be also created on the go as needed by taking into account situated and contextually meanings. This dynamic semantic system does not contradict the assumption of an internalized schema of a lexical concept. In contrast, these schemas may scaffold the creation of new lexical internalizations. Schemas seem to be important conceptual building blocks as many lexical concepts are not acquired by a set of instances such as belief, poetic, attitude or language faculty. The generation of semantic primitives for sentence processing may rely on particular rules and principles. To explore these semantic mechanisms may be a more fruitful approach than trying to establish a complete list of semantic primitives.

Furthermore, we can assume that these generative semantic rules are innate and specific to Homo sapiens. It is quite speculative, but it seems plausible to postulate a parallelism between semantic and syntactic rules. Both types of rules appear to be fundamentally different, but future research may reveal that those cognitive rules and principles are innate and universal, which apply not only for the linguistic domain. The long-term goal as understood here is to characterize these human-specific cognitive mechanisms, from which domain-specific rules such as semantics and syntax are derived.

For example, Jackendoff (1983, 1990, 1992) considers conceptual semantics as part of thought and not as part of language. He identifies a set of universal semantic categories such as Thing, Event, State, Path, Place, Property, Time, or Amount. They can be considered as conceptual syntactic categories generating the sematic contents of sentences. By comparing the syntactic (12a) and conceptual structure (12b) of a sentence (Jackendoff 1992), the NPs correspond to Thing, the verb to Event, and the PP to Path:

(12) a. [$_{S}$[$_{NP}$ Bill] [$_{V}$ went] [$_{PP}$[$_{P}$ into] [$_{NP}$ the house]]]]
b. [$_{Event}$ GO([$_{Thing}$ Bill] [$_{Path}$ TO ([$_{Place}$ IN ([$_{Thing}$HOUSE])])])]

Similar, the concept Cause is a semantic primitive as it can account for the different verb meanings expressed in (13a–d):

(13) a. Mary lifted her son.
b. Mary gave the toy to her son.
c. Mary persuaded her son to play hockey.
d. Mary killed the bee.

It is certainly questionable to what extent semantic primitives are primitive, as mentioned before, and they change in light of specific contextual triggers. However, Jackendoff's framework shows how the interface between conceptual and linguistic (semantic-syntactic) structures can be designed. Most semantic primitives proposed in linguistics are abstract concept, which are mostly so far immune to empirical evaluation.

In sum, lexical concepts may have a compositional structure and meanings are composed of smaller parts, across different sensory information and lexical domains. But it should be emphasized that here we do not make any claims about the cognitive mechanism underlying these compositions. The automatic and effortless lexical acquisition process shows that these processes are not conscious or intentional inferences. What appear to rely on a genetically endowed capacity are the underlying cognitive mechanisms of concept development (Fodor 1998). Assuming that we would be able to trace back the lexical conceptual acquisition process, it might turn out that we arrive at a point where basic information (semantic primitives) will not be learned. These primitives may be given to the learner genetically along with the characteristics of human brain development (see also Pinker and Levin 1991).

It is intuitively plausible what we have discussed so far: Speakers have access to lexical information, which is without any doubt very rich in terms of size and facets. The complexity of the topic is reflected by a wide range of different approaches, of which we can address here in the present context only a few, mostly relevant for the link between cognitive and neural structures. Research on aphasia and dementia revealed that single patients show comprehension disorders with respect to particular semantic categories or sensory-specific information (e.g., Warrington and Shallice 1984; Goodglass et al. 1986; Hillert 1992). The evidence reported, provides a hint that the information of the semantic lexicon is not as systematically organized as most linguistic theories proposed. In light of selective mechanisms, even the traditional debate whether the semantic structures *per se* are impaired or the lexical semantic access mechanisms seems not to be relevant anymore. The data indicate that some domains can be selectively impaired, but they do not provide positive evidence for a particular lexical model. In fact, it can be argued that the structure of the conceptual lexicon can significantly vary from person to person. Thus, how the brain organizes lexical information seems to be more compatible with distributed representations rather than in the form of a mental dictionary. This approach is also supported by neuroimaging and electrophysiological evidence that for example action words related to face, arm, or leg elicites in addition to temporal lobe and inferior frontal gyrus (IFG) activity specific somatotopic activation patterns along the motor strip (e.g., Hauk et al. 2004). Those data indicate that certain meaning aspects of action words are accessed in the proximity of motor experience. But the data also indicate that the meaning of a word is not a static concept but will be accessed (and probably constructed) at different cortical sites. In particular for abstract lexical meanings or complex expressions, the access points may vary from person to person depending on discourse context, episodic experience, and situational factors.

7.2 Mental Space

Let us briefly return to the 1970th when semantic network models proposed by cognitive psychologist tried to draw a more realistic picture of how words might be stored in the brain. In particular, those psychological models described single word meanings set- and/or net-theoretically (e.g., Rumelhart et al. 1972; Anderson and Bower 1973; Glass and Holyoak 1974) or according to typicality and basic concepts (Labov 1973; Rosch 1978; Mervis and Rosch 1981; see also Wittgenstein 1953). Those models address, however, only partially the significant role of contextual information in processing word meanings, but emphasize the dynamics of processing lexical meanings (see Johnson-Laird 1987). A more recent approach, appears to be an extension of these ideas by assuming that word meanings should be considered as internal states or also called mental states. Elman (2004, 2011) uses a simple recurrent network to simulate the mental states at the level of the hidden layer. A mental state depends on its own state at time point t to feed into the state at time $t + 1$ and the external stimulus or input (here words). This simple recurrent network learns the strength of weights between words from scratch. Words are presented in form of binary vectors and a simple learning algorithm modifies the connection weights (Rumehart et al. 1986). If the training input set is large and complex enough, the net learns the context-contingent dependencies, which computes the probability of succeeding words (and excludes ungrammatical words).

Elman (2011) provides the following example: *For example, after learning, and given the test sentence "The girl ate the...," the network will not predict a single word, but all possible words that are sensible in this context, given the language sample it has experienced. Thus, it might predict sandwich, taco, cookie, and other edible nominals. Words that are either ungrammatical (e.g., verbs) or semantically or pragmatically inappropriate (e.g., rock) will not be predicted.* The training of the net clusters similarity structures between words and in turns out that the net treats nouns independent of verbs. Again, within the cluster of nouns, the net differentiates for example between the semantics of animate vs. inanimate, animals vs. humans, large vs. small animals, food vs. breakables; within the cluster of verbs for example between only intransitive vs. possibly transitive vs. only transitive. Moreover, the verb has a particular meaning in a specific sentence context. Thus, in considering a particular situated context, the verb meaning assigns specific arguments and adjuncts. Based on lexical input, the net predicts the thematic grid of the verb's meaning. For example, the net predicts that the sentence *The butcher uses a saw to cut ...* will be completed by the word *meat* and the sentence *A person uses a saw to cut ...* by the word *tree*. Elman refers to two related strategies the net is learning to make these predictions. First, the net learns the lexical semantic attributes of the words including hierarchical relationships as mentioned above. Second, the net learns the syntagmatic (grammatical) knowledge of argument-adjunct-verb interactions encoded in the weights between the words. Thus, a word's impact on the internal (mental) state combined with prior context produces predictions about the subsequent syntactic class, that is, about the phrase structure. Figure 7.1 illustrates

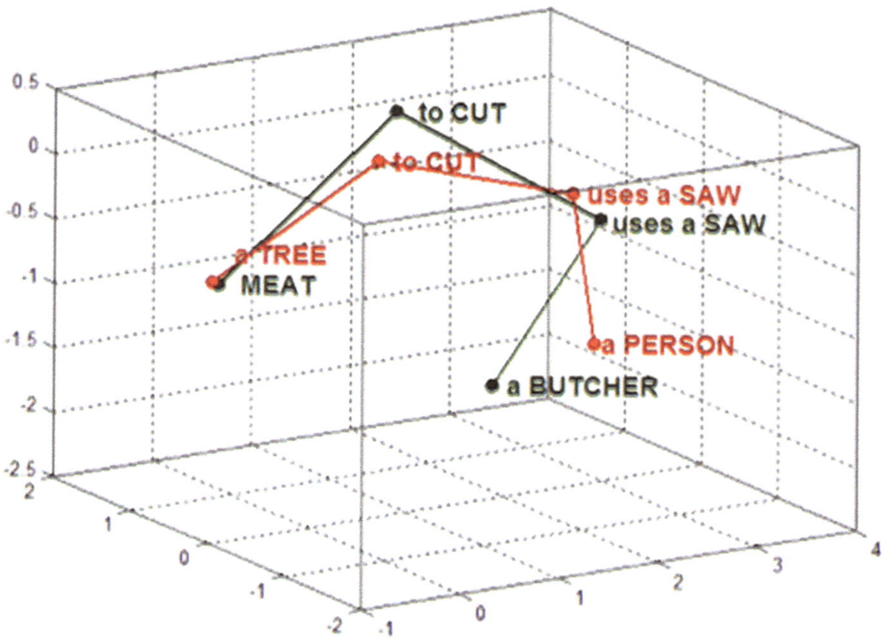

Fig. 7.1 3D trajectories for both sentences shown result from the internal states of each word in the hidden layer of a simple recurrent network. Each word encodes expectancies of what will follow. The state at *cut* differs in both sentences reflecting different predictions regarding the semantic role Patient to be followed. (Adapted, Elman 2011; © John Benjamins Publishing Co.)

the trajectories of the net's internal state space when two sentences with the verb *cut* are processed. The meanings of both sentences are closely related but not identical.

As emphasized also by Elman, this is a very basic lexical model but shows an alternative approach to the traditional view of lexical representations. Here, the lexicon is empty as the meaning of words is dynamically determined by the specific linguistic context. Of course, a full model must considers all factors relevant for generating the intended meaning. In particular, context refers here also to discourse-level information, sensory-motoric experiences, or complex event knowledge, which is not verbally expressed. Moreover, the net must be able to bypass expected patterns as it is often applies to new metaphoric expressions, puns, or rhetoric formulations. Thus, the learned dynamics of a neural network itself may be not sufficient to generate meanings, but must use complex inferences to achieve the intended mental state.

Related but different approaches are those statistical models, which infer lexical meanings from large text corpora. For instance, latent semantic analysis (LSA) determines the meaning of a word by considering all the information linked to this work in the text corpora (Landauer et al. 1998). LSA represents the words in a very high (e.g., 50–1,000) dimensional semantic space. Thereby, the analysis does not simply sum up the pairwise or tuple-wise co-occurrence of words, but detailed lexical occurrence patterns over sentences or sections. In the hyperspace analogue to language model (HAL), semantics is generated by moving an *n*-word (e.g., 10) window across phrases and counting the number of words between any two words (Lund and Burgess 1996). The semantic strength between two words is inversely incremented with the distance between these two words. As in LSA, the semantic similarity between two words is calculated by the cosine of the angle between their vectors. However, the application of large text corpora for the generation of word meanings reflects the contextual usage of a word but it does not inform about the cognitive or neural representation of semantics. What kind of semantic information do we store in long-term memory to construct the meaning to be intended? At present, most evidence speaks for a distributed semantic network. Future work may provide a more precise picture of how meanings are acquired in childhood and how meanings are represented and constructed that overlapping mental states among speakers are the foundation for successful communication.

References

Anderson, J. R., & Bower, G. H. (1973). *Human associative memory*. Washington, DC: Winston and Sons.

Croft, W. (1993). The role of domains in the interpretation of metaphors and metonymies. *Cognitive Linguistics, 4*(4), 335–370.

Elman, J. L. (2004). An alternative view of the mental lexicon. *Trends in Cognitive Sciences, 8*(7), 301–306.

Elman, J. L. (2011). Lexical knowledge without a lexicon? *Mental Lexicon, 6*(1), 1–33.

Evans, V. (2004). *The structure of time: Language, meaning and temporal cognition*. Amsterdam: John Benjamins.

Fauconnier, G. (1997). *Mappings in thought and language*. Cambridge: Cambridge University Press.

Fodor, J. A. (1975) *The language of thought*. Cambridge, MA: Harvard University Press.

Fodor, J. A. (1987). *Psychosemantics. The problem of meaning in the philosophy of mind*. Cambridge: MIT Press.

Fodor, J. A. (1998). *Concepts: Where cognitive science went wrong*. Gloucestershire: Clarendon Press.

Glass, A. L., & Holyoak, K. (1974). The effect of some and all on reaction time for semantic decisions. *Memory & Cognition, 2*(3), 436–440.

Goodglass, H., Wingfield, A., Hyde, M. R., & Theurkauf, J. C. (1986). Category specific dissociations in naming and recognition by aphasic patients. *Cortex; a journal devoted to the study of the nervous system and behavior, 22*(1), 87–102.

Hauk, O., Johnsrude, I., & Pulvermüller, F. (2004). Somatotopic representation of action words in the motor and premotor cortex. *Neuron, 41*, 301–307.

Hillert, D. (1992). Lexical semantics and aphasia: A state-of-the-art review. *Journal of Neurolinguistics, 7*(1), 1–43.

Jackendoff, R. (1983). *Semantics and cognition*. Cambridge: MIT Press.

Jackendoff, R. (1990). *Semantic structures*. Cambridge: MIT Press.

Jackendoff, R. (1992). *Languages of the mind: Essays on mental representation*. Cambridge: MIT Press.

Johnson-Laird, P. N. (1987). The mental representation of the meaning of words. *Cognition, 25*, 189–211.

Labov, W. (1973). The boundaries of words and their meanings. In C. J. Bailey & R. Shuy (Eds.), *New ways of analyzing variation.* (In English) (pp. 340–373). Washington, DC: Georgetown University Press.

Lakoff, G. (1987). *Women, fire, and dangerous things: What categories reveal about the mind.* Chicago: University of Chicago Press.

Landauer, T. K., Foltz, P. W., & Laham, D. (1998). Introduction to latent semantic analysis. *Discourse Processes, 25,* 259–284.

Lund, K., & Burgess, C. (1996). Producing high-dimensional semantic spaces from lexical co-occurrence. *Behavior Research Methods, Instruments & Computers, 28*(2), 203–208.

Mervis, C. B., & Rosch, E. (1981). Categorization of natural objects. *Annual Review of Psychology, 32*(1), 89–115. doi:10.1146/annurev.ps.32.020181.000513.

Pinker, S., & Levin, B. (1991). *Lexical and conceptual semantics.* Cambridge: The MIT Press.

Rosch, E. (1978). Principles of categorization. In E. Rosch & B. Lloyd (Eds.), *Cognition and categorisation* (pp. 27–48). Hillsdale: Erlhaum.

Rumelhart, D. E., Lindsay, P. H., & Norman, D. A. (1972). A process model for long-term memory. In E. Tulving & W. Donaldson (Eds.), *Organization and memory.* New York: Academic Press.

Rumelhart, D. E., McClelland, J. L., & PDP Research Group, C. (Eds.) (1986). *Parallel distributed processing: Explorations in the microstructure of cognition, vol. 1: Foundations.* Cambridge: Bradford.

Turner, M. B., & Fauconnier, G. (1995). Conceptual integration and formal expression. *Metaphor and Symbolic Activity,* 183–203.

Van Langacker, D. (1987). Comprehension of familiar phrases by left- but not by right-hemisphere damaged patients. *Brain and Language, 32*(2), 265–277.

Warrington, E. K., & Shallice, T. (1984). Category specific semantic impairments. *Brain: a journal of neurology, 107,* 829–854.

Wittgenstein, L. (1953). *Philosophische Untersuchungen [German].* Frankfurt a. M.: Suhrkamp.

Chapter 8
Figurative Language

8.1 Lexical Dark Matters

Mainstream linguistics tends to consider the structure of figurative language, not as relevant or as odd unimportant cases for the formulation of a linguistic theory. Particularly motivated by empirical research in psycho- and neurolinguistics, different models were however discussed about the computational differences between literal and non-literal languages, but in general, figurative language still is considered as a poor relative of standard linguistic grammar (but see for example Langacker 1987; Fauconnier 1997; Jackendoff 1997; Hillert 2008). Here, we are convinced that any linguistic theory should not classify figurative language as problematic, idiosyncratic or as irregular infrequent cases, but may open the door to new insights about the human language system. The use of figurative language is actually highly common in everyday conversation, although dictionaries typically do not document the variety and frequency of different non-literal expressions. Jackendoff (1997) estimates 25,000 idiomatic expressions alone, but idioms are only a fraction of the broad range of tropes used on a daily basis.

By using text mining methods, it was found that the size of written English words nearly doubled over the past century to more than one million words (Michel et al. 2011). From 1900 to 1950 the increase was estimated to be around 50,000; however, from 1950 to 2000 there was a jump of about 500,000 words. At the same time, well-known dictionaries such as the Oxford English dictionary (Stevenson 1993) or the Merriam-Webster Unabridged dictionary (Gove 1993), both document only about 25 % of those words.

The gap between dictionary documentation and usage of written words increased significantly. Michel and colleagues estimate that 52 % of the words used in English books are lexical dark matters, undocumented in standard references such as the Oxford English dictionary. These lexical dark matters include, in particular, infrequent words, but particularities such as proper nouns or compounds were excluded from the analyses. The usage of words is relatively dynamic, words are born and die, and some words have a long, some a short life-cycle. We assume that large portions of the lexical dark matters belong to the category of tropes, a lexical category more

D. Hillert, *The Nature of Language*, DOI 10.1007/978-1-4939-0609-3_8,

dynamic and creative than words with static references expressing literal meanings. It is difficult to estimate the amount of tropes used in a language, but it can make up 50 % or more of the words used in a particular language, even if regional differences are not considered. Wikipedia is listing 118 different types of tropes and the list is far from complete. Thus, neurolinguists should be obliged to extend their field of research to tropes to draw a more realistic picture about the human language circuits.

To begin with, let us first look at some typical variations of figurative language. Often, we are not aware of the origin of a conventionalized figurative expression, but we learned that particular sayings are associated with sophisticated meanings otherwise difficult to express. Often these figures of speech have rhetorical purposes, communicate particular emotional states, create images or other cross-sensory experiences and have a unique typology regarding prosody, intonation, word choice, grammar, etc. However, because each single figurative expression is idiosyncratic and has its own uniqueness, different classes of figurative expressions often share particular characteristics. Thus, the examples provided serve only as guidelines. In metaphors, a meaning is created that blends two different mental spaces otherwise not connected: *The park is the green lung of the city.* An idiom is a conventionalized fixed expression. Its decomposition can be often literally interpreted, as in *The teacher spilled the beans*, but typically the figurative reading is more frequent. The degree of grammatical frozenness varies from expression to expression.

A saying that expresses some sort of truth experience is called a proverb: *Too many cooks spoil the broth.* In the case of a metonym, the meaning is shifted within the same mental space as in the case of *Washington said ....* A synecdoche is a type of metonymy, in which the parts represent the whole: *The jaw injured the surfer.* The properties of a particular mental space are exemplified in a simile, as in *He fought like a tiger.* A hyperbole involves an exaggeration, such as *The Brazilian summer kills us*. In opposite, a meiosis or litotes involves an understatement: *It just got a little bit out of hand,* when commenting on a wild party. In the case of sarcasm or irony the opposite meaning will be expressed: *I have to work all weekend.* Response: *How nice!* An oxymoron implies a paradox or contradiction as in the case of *mad wisdom*. Finally, puns use often a word play involving ambiguities or contradictions: *A rule of grammar: double negatives are a no-no.*

8.2 Idioms and Metaphors

Idioms are a subset of a broad range of different types of fixed expressions such as collocations, titles and names, or verb particle constructions (Jackendoff 1997, 2002; Hillert and Ackermann 2002). They are idiosyncratic as they are often violating morphosyntactic rules. An idiom may be stored as a single unit in the mental lexicon, because its figurative meaning cannot be derived from the individual word meanings it is composed of. Spoken comprehension of an ambiguous sentence such as *Mary breaks the ice* involves several parsing steps. Initially, the parser segments and merges incoming speech and generates a hierarchical syntactic structure between the lexical elements of the sentence. Regarding a literal reading of example (14), the verb assigns the Θ-role Agent (AG) to the subject noun phrase (NP) and reserves an argument slot for the Θ-role Theme (TH). The subject NP and the following verb phrase (VP) need to be merged to generate a sentence structure.

(14)

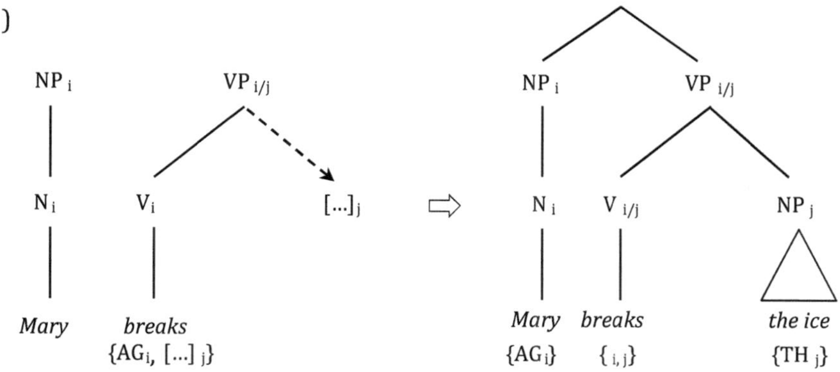

In turn, the Θ-role TH will be activated and the parser's output will be mapped onto a semantic representation. For different possible reasons such as the bias of a pragmatic context and/or the high familiarity of an idiomatic expression, the parser computes at the idiom recognition point, a non-compositional alternative reading. If this occurs at a very early access stage, the parser needs only to suppress the thematic slots, already activated to assign the verb an intransitive function (15). In general, different parsing strategies can apply, depending on the onset cue, contextual bias and/or frequency. Which factors or combination of factors determine access to a figurative reading, is still controversially discussed.

(15)

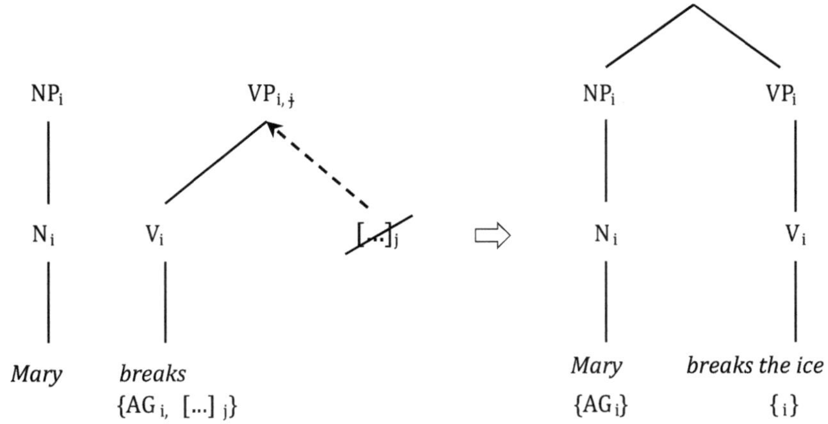

Idioms vary in their degree of morphosyntactic flexibility (e.g., Katz and Postal 1964; Fraser 1970; Newmeyer 1974). Whether certain syntactic operations can apply to a particular idiomatic expression seems to depend, in many cases, on pragmatic functions.

(16) a. [The bucket was kicked]$_I$ [by Sam]$_J$
=> *[Died]$_I$[by Sam]$_J$

b. [The beans]$_I$ [were spilled]$_J$ [by Mary]$_K$
=> [The information]$_I$ [was revealed]$_J$[by Mary]$_K$

For example, a figurative reading of the passive phrase in (16a) sounds odd, because the focus *the bucket* is not corresponding to any discourse entity. However, the passive phrase (16b) preserves a literal and a figurative interpretation, but in the case of (16a) all lexical elements of the VP refer as a whole to a single concept. Again, the intransitive verb *seem* can modify very often an idiomatic expression. Example (17) indicates that an NP movement is involved.

(17) The proposal does not hold water.
[The proposal$_i$] seems [_$_i$ not to hold water]

In this example, however, a literal interpretation is less plausible, but the parser still needs to revise its initial (de)compositional analysis. While morphosyntactic rules apply at the surface structure of a sentence, how is it possible to license units larger than X^0 (terminal nodes), if syntactic movements can apply?[1]

Two main proposals are discussed. In Chomsky's (1981, note 94) version, verbal idioms are treated as lexical verbs with internal structure (18). In Jackendoff's approach (1997), the complete idiomatic expression is indexed to the lexical conceptual level. As already discussed in (16), subject and direct object in *kick the bucket* are not stipulated at the semantic level (19a, b).[2]

(18) a. [$_V$ [$_{VP}$ [$_V$ kick][$_{NP}$ the bucket]]]

b.

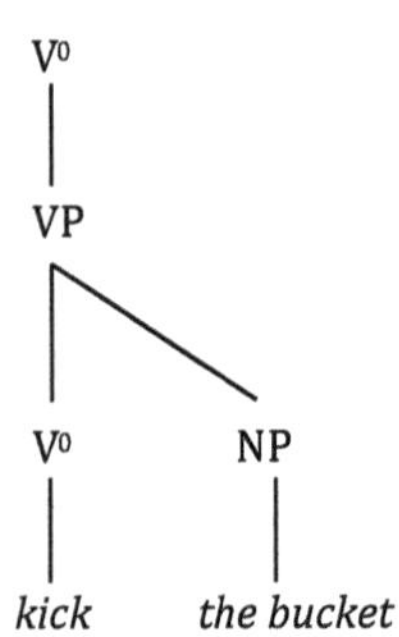

[1] In generative grammar, the surface structure (S-structure) of a sentence is derived from its deep structure (D-structure) via syntactic movements, traditionally called 'transformations' (Chomsky 1981). In X-bar (X') theory X^0 is the head or terminal node of the phrase, which is also called zero projection (e.g., Chomsky 1986; Di Sciullo and Williams 1987). Its value ranges over at least the lexical categories N (noun), A (adjective), V (verb) and P (proposition).

[2] Halle and Marantz (1994) postulated an additional semantic level of structural meanings to account for the observations that *kick the bucket* cannot mean *die,* because the verb phrase subcategorizes a direct object.

(19) a. $[_{VP}\ [_{V}\ \text{kick}][\ _{NP}\ \text{the bucket}]_{j}]_{x} \Leftrightarrow [_{Event}\ \text{DIE}\ ([\ ]_{i}[\ ]_{j})]_{x}$

b. 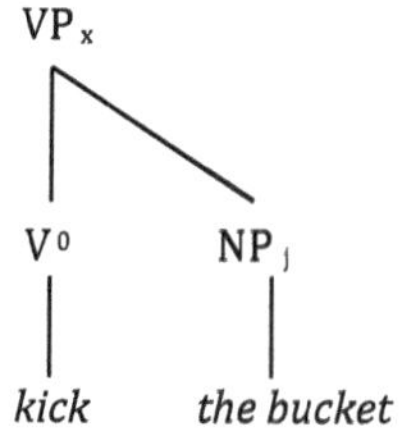

It is obvious that semantic information plays an important role in describing the syntactic flexibility of idioms (Wasow et al. 1983). However, there is often no direct correlation between the lexical semantic and syntactic levels, because many non-compositional idioms are, to some extent, syntactically flexible. As discussed below, a formal syntactic approach seems to be most suitable for analyzing the linguistic heterogeneity of idiomatic expressions and for predicting variance of computational complexity.

Herewith an idiom account is proposed, that considers individual syntactic flexibility in addition to preserving the figurative meaning at the syntactic level (19). This approach is, to some extent, comparable with Van Gestel's (1995) "en bloc insertion" account. Grammatical features pose constraints on the syntactic flexibility of idiomatic compounds (19a) or phrases (19b–c). They are introduced as negative features (inhibitors) at the head-level of the idiomatic structure that is co-indexed with the relevant semantic concepts (e.g. PLU, plural; Gen, genitive; PASS, Passive; TOP, topicalization; SGL, singular).

(20)

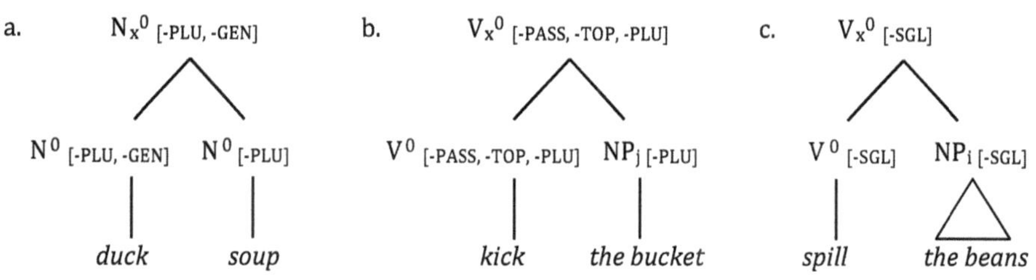

(21) a. *Duck soup*: $[_{State}\ \text{EASY}]_{x}$

b. *Kick the bucket*: $[_{Event}\text{DIE}\ ([\ ]_{i}\ [\ ]_{j})]_{x}$

c. *Spill the beans:* $[_{Event}\ \text{REVEAL}\ ([\ ]_{i}[\text{SECRET}]_{j})]_{x}$

Because of their idiosyncratic nature, it is difficult to define what idioms are, in particular as we are usually unaware of their origin. One might say that idiomatic meanings can be considered as semantically and syntactically frozen creative extensions of literal meanings. As we have seen, idiomatic structures require access to alternative parsing structures and may therefore involve higher computational costs or a higher cognitive demand than a typical literal parse without a figurative reading. The second category of figurative speech refers to metaphors which we discuss below in more detail.

(22) a. The athlete kicked the sport's bucket as he has been fired.
b. He's burning the midnight oil from both ends.

Metaphoric extensions are sometimes used to modify the meaning of a frozen idiomatic expression as illustrated in (22). A novel meaning will be created, which is more specific than the standard meaning of the idiom.

Novel metaphoric expressions do not require alternative parsing strategies for comprehension, as it is sometimes required for idiomatic expressions, but semantic features will be interpreted in an atypical conceptual space. For instance, in the relatively non-conventionalized novel metaphoric expression *The botanical garden is the green lung of the city,* the concept *botanical garden* receives new semantic attributes from the concept LUNG (+breath), which allows a comparison between the function of a lung and a botanical garden. The extension of the core meaning of a lexical concept GARDEN by the semantic attributes of the core meaning of the lexical concept LUNG is illustrated in (23). Thus, metaphoric interpretations may occur post-syntactically at a pragmatic level (e.g. Lakoff and Johnson 1980; Gibbs 1994; Fauconnier 1985).

Some theorists may argue that any meaning is metaphoric, as symbolic or mental labels such as words, expressions or sentences never match the external object or event, but provide symbolic approximations (see Quine 1960). Indeed, common words such as *lady, king, priest* or *person* may have originated in metaphoric meanings before they become semantically frozen, mainly because of a high frequency of usage. To understand the language system, it is irrelevant *what* has been said, but more important *how* it has been said. In addition to a common usage of words, speakers are able to use this semantic knowledge as a base for creating different strategies for expressing meanings. Thus, different kinds of conceptual strategies will be used to express meanings and intentions. Metaphors map thoughts to language in a relatively direct fashion. Although nonfigurative language also reflects conceptual strategies, they are often intermingled with over-learned linguistic routines.

(23)

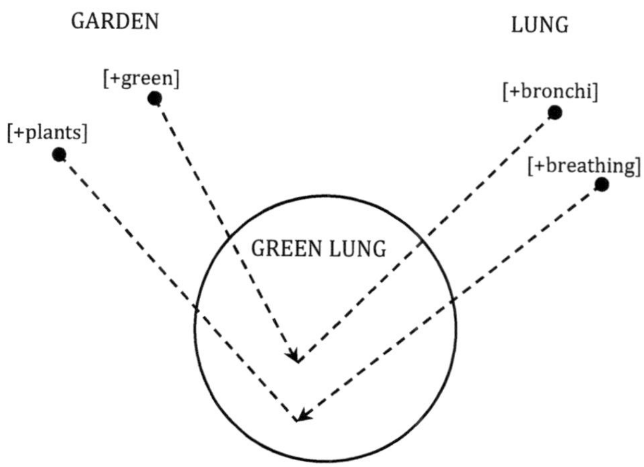

In concluding, we would like to emphasize that novel metaphors and frozen expressions may represent two endpoints of a semantic continuum, in which varying grades of figurativeness are processed. Those computations may not be specific to figurative meanings, but may be highly automatic or constructive as in the case of non-literal meanings. In particular, speech errors found in figures of speech show similar linguistically and/or conceptually motivated patterns as in non-figurative speech. For example, the less frequent old-fashioned idiomatic expression *The shit hits the fan* has been modified by the exchange of the nouns (24a); similarly, *Wolf in sheep's clothing*, a less common idiom with a Biblical origin, has been phonologically alternated, supported in addition by the common adjective–noun combination *cheap clothing* (24b); again, in the next example the verb has been replaced and the new phrase is plausible, if you know the original idiom (23c).

(24) a. The fan is hitting the roof
The shit hits the fan.

b. Wolf in cheap clothing
Wolf in sheep's clothing

c. I wouldn't eat that with aten-foot pole.
I wouldn't touch that with a ten -foot pole.

The blended figurative phrases mentioned above indicate compositional elements involved in production. However, based on the large variety of figurative phrases, particularly in the class of metaphors, it is difficult to design a unitary theory of metaphors. Most approaches differentiate between different classes of metaphors based on their computational differences; some behave like idioms as they are highly conventionalized, others might be used only once or presumably only by chance, or certain novel constructions might be used more commonly. It appears thus that a unitary theory would not be a suitable approach for the description of metaphors. The large variety of different metaphors requires analyzing each expression separately, as even the attempt to establish different classes seems to cover only a small portion of metaphors used in everyday speech. After all, metaphors are highly useful access points for understanding individual thoughts.

References

Chomsky, N. (1981). *Lectures on government and binding*. Dordrecht: Foris.

Fauconnier, G. (1985). *Mental spaces*. Cambridge: MIT Press.

Fauconnier, G. (1997). *Mappings in thought and language*. Cambridge: Cambridge University Press.

Fraser, B. (1970). Idioms within a transformational grammar. *Foundations of Language, 6,* 22–42.

Gibbs, R. W. (1994). *The poetics of mind: Figurative thought, language, and understanding*. Cambridge: Cambridge University Press.

Gove, P. B. (1993). *Webster's third new international dictionary*. Springfield: Merriam Webster.

Hillert, D. G. (2008). On idioms: Cornerstones of a neurological model of language processing. *Journal of Cognitive Science, 9,* 193–233.

Hillert, D., & Ackerman, F. (2002). Accessing and parsing phrasal predicates. In N. Dehé, R. Jackendoff, A. McIntryre, & S. Urban (Eds.), *Verb-particle explorations* (pp. 289–313). Berlin: Mouton de Gruyter.

Jackendoff, R. (1997). *The architecture of the language faculty*. Cambridge: MIT Press.

Jackendoff, R. (2002). English particle constructions, the lexicon, and the autonomy of syntax. In N. Dehé, R. Jackendoff, A. McIntryre, & S. Urban (Eds.), *Verb-particle explorations* (pp. 67–94). Berlin: Mouton de Gruyter.

Katz, J. J., & Postal, P. M. (1964). *An integrated theory of linguistic descriptions*. Cambridge: MIT Press.

Lakoff, G., & Johnson, M. (1980). *Metaphors we live by*. Chicago: University of Chicago Press.

Langacker, R. W. (1987). *Foundations of cognitive grammar: Theoretical prerequisites* (Vol. 1). Stanford: Stanford University Press.

Michel, J. B., Shen, Y. K., Aiden, A. P., Veres, A., Gray, M. K., Google Books Team, Pickett, J. P., Hoiberg, D., Clancy, D., Norvig, P., Orwant, J., Pinker, S., Nowak, M. A., & Aiden, E. L. (2011). Quantitative analysis of culture using millions of digitized books. *Science, 331*(6014), 176–182.

Newmeyer, F. J. (1974). The regularity of idiom behavior. *Lingua, 34*(4), 327–342.

Quine, W. V. O. (1960). *Word & object*. Cambridge: MIT Press.

Stevenson, A. (1993). *Oxford English dictionary*. Oxford: Oxford University Press.

Van Gestel, F. (1995). En bloc insertion. In M. Everaert, E.-J. van der Linden, A. Schenk, & R. Schreuder (Eds.), *Idioms: Structural and psychological perspectives*. Hillsdale: Lawrence Erlbaum.

Wasow, T., Sag, I., & Nunberg, G. (1983). Idioms: An interim report. In S. Hattoru & K. Inoue (Eds.), *Proceedings of the XIIIth international congress of linguistics* (pp. 102–115). Tokyo: Nippon Toshi Center.

Part III
Circuits

Chapter 9
Generating Sentences

9.1 Structural Complexity

The human brain has a predisposition for organizing perceptually discrete units into sequences that are hierarchically organized. These structures follow combinatorial principles or syntactic rules and are most prominent in language and music. Here, we focus mainly on the linguistic capacity as a gate to those underlying cognitive principles and neural correlates that seem to be universal to human nature. Those universal principles underlie typological varieties of natural languages that rise and fall along with the presence of culture entities.[1] To comprehend a sentence (well-formed or not), speakers intrinsically activate not only long-term syntactic knowledge, but perform processes or operations on the base of this knowledge for unifying units according to specific syntactic rules of their language.

This dual approach applicable not only to syntax but to most cognitive skills acquired can be imagined as a specialized syntactic network. Representations and computations in such a neural net operate in the same dimension but may recruit different kind of processes. While representations may vary in their degree of activation levels, computations may differ with respect to their operational costs. Before neuroimaging tools were commonly applied in research, evidence relied mostly on the analyses of agrammatic speech and syntactic comprehension patterns found in aphasic lesion studies (e.g., Hillert 1990, 1994; Goodglass 1993). Thus, it has been assumed that Broca's area (BA 44/45 equiv. to F3op/F3t) is a cortical region specialized for syntactic processing.[2] Along with the development of neuroimaging tools, improved psycholinguistic online methods and a more fine-grained theoretical framework, it was possible providing evidence that this conclusion is an oversimplification. Moreover, many empirical data did not support the claim that Broca's area is exclusively involved in syntactic operations.

---

[1] In Sept. 2012, "Ethnologue" classifies 6,909 distinct languages (Lewis, ed. 2009).

[2] Typically, Broca's area involves BA 44 & 45, but some studies include also BA 47. In addition, the terms *pars opercularis* (F3op) and *pars triangularis* (F3t) are used which are not respectively coextensive with BA 44 and BA 45.

D. Hillert, *The Nature of Language*, DOI 10.1007/978-1-4939-0609-3_9,

At this point, some methodological concerns should be mentioned. Because of the application of functional magnetic resonance imaging (fMRI) we are able to draw a more precise picture about the neural substrates that underlying specific cognitive computations. We believe that in the linguistic domain most fMRI data may provide in general useful information on the correlated neural activity although a few colleagues addressed this line of empirical research as voodoo science referring to questionable correlation methods (Vul et al. 2009). Indeed, as any empirical method fMRI has methodological issues and meaningful criticism may help to advance the neuroimaging field. It seems critical that the evaluation of MRI data relies typically on group analyses and do not report statistics of individual cases. However, this is a general methodological flaw of behavioral studies (including psycholinguistic research), but often significant effects can be only reported at the group level. Furthermore, fMRI results are typically reported as contrasts, that is, condition A is stronger active as compared to condition B. This however does not imply that B is not active at all. The relation between A and B is reported as a significant effect of A. Keeping these points in mind, let us return to fMRI evidence of syntactic computations at the sentence level.

First, the application of fMRI shows that the activation of Broca's area is not restricted to particular types of syntactic structures, but is also involved in a range of different phonological and lexical semantic operations (e.g., Burton 2001; Bookheimer 2002; Hillert and Burač as 2009). Second, Broca's area will be also activated in other domains such as motor or rhythmic activations (e.g., Iacoboni et al. 1999; Herdener et al. 2012). In considering these different types of evidence, it is apparent that the computations performed by Broca's area are not exclusively involved in sequencing hierarchical structures of lexical items but also in other cognitive operations involving the generation and linearization of structural hierarchies. It needs to be considered that a lesion in Broca's area is neither sufficient nor necessary to cause disorders of (morpho-)syntactic processing (Dronkers et al. 1994). For example, access to syntactic information is not *per se* impaired in Broca's aphasic patients as they are able to judge the grammaticality of certain sentence structures (Linebarger 1983). Finally, it is important to emphasize that syntactic operations typically involve not only Broca's area, but also left and right temporal lobe areas (BA 21, 22), as revealed by different fMRI studies (e.g., Mazoyer et al. 1993; Just et al. 1996; Embick et al. 2002; Meyer et al. 2000; Moro et al. 2001). For example, in Mazoyer et al.'s study bilateral activities of the temporal poles were found for listening to stories compared to words. However, often these studies are not designed to differentiate between semantic and syntactic computations at the sentence level.

In general, we still are not informed whether specific structural subroutines can be associated with subregions of Broca's area. The reference to a few studies serves in this context as representative examples. In Dapretto and Bookheimer's (1999) fMRI study subjects were asked to decide whether two (auditorily presented) sentences were the same or different. The syntactic condition consisted of an active

(25a) and a passive sentence (25b); in the semantic condition, the sentence pair differed by a single word (26a, b).[3]

(25) a. The policeman arrested the thief.
 b. The thief was arrested by the policeman.

(26) a. The lawyer questioned the attorney.
 b. The driver questioned the attorney.

Significant activations were found in BA 44 for syntactic operations (in comparison to semantic operations or to the rest condition); however, semantic operations seem to be associated with BA 47 (semantics minus syntax). The opercular portion of Broca's area apparently is specifically active when the computations require syntactic (structure-driven) operations rather than semantic (content-driven) operations.

In another fMRI study by Kang et al. (1999), the stimulus material consisted of verb phrases (VPs); in addition to the correct baseline condition (27a), syntactic (27b) and semantic (26c) violations were visually presented.

(27) a. drove cars
 b. *forgot made
 c. *wrote beers

The findings seem to confirm Dapretto and Bockheimer's results as such that both types of violations recruited activations in BAs 44 and 45 as compared to the baseline whereas greater activity was calculated for the syntactic violation in BA 44. This study is however also an example of how difficult it is to disentangle different types of linguistic levels. Both types of violations mentioned in (27a) and (27b) may also involve not only syntactic but also semantic processes to generate meanings of apparently unrelated words (e.g., *wrote [about] beer; [he] wrote beer*). How speakers compensate for linguistic violations is a gray area and probably depends on preference strategies. Thus, phrasal violation studies may reflect the outcome of different degrees of linguistic violations rather than valuable date about the difference between syntactic and semantic computations. Interestingly, BA 44 seems to be stronger engaged than BA 45 if the violation seems to be stronger.

By varying the degree of syntactic (or better sentence) complexity, it is in most cases difficult to decide whether the effects found are due to syntax alone or to other cognitive factors such as computational costs associated with the length or complexity of a sentence. For example, in using positron emission tomography (PET), Stromswold et al. (1996) found in a semantic plausibility task for center-embedded sentences stronger BA 44 activities as compared to right-branching relative-clause

[3] In this context, a general picture about some relevant findings will be drawn and details of methodological differences among different studies such as design issues (e.g., block vs. event-related design; modality of presentation), which usually modify the final results and thus may lead in some cases to different conclusions.

sentences and assumed a computational load effect (see Just et al. 1996). In a follow-up fMRI study by Caplan et al. (2008), comparable center-embedded (28a) and right-branching structures (28b), which represent both object-extracted sentences, were examined in context of different task conditions (sentence verification, plausibility judgment or non-word detection task) to examine task-specific effects on syntactic processing.

(28) a. The juicei that the child spilled ti stained the rug.

b. The child spilled the juicei that ti stained the rug.

Comparable findings were found, and based on their study design, Caplan et al. conclude that the activation in the left IFG (Broca's area) is task-independent. While this study shows that the effect cannot be reduced to task-specific factors, it remains open whether these findings are caused by variations of syntactic or computational complexity, or by a combination of both factors.

Numerous studies were designed to examine whether an increase of computational costs is related to an increase of working memory (WM) load or of syntactic complexity. The research focused on the comparison between object-extracted noun phrases (NPs) as compared with subject-extracted NPs (e.g., Caplan et al. 1998). The assignment of syntactic structures and θ-roles involves higher computational costs in object-extracted NPs than in subject-extracted NPs. In contrast to (28), in Hebrew topicalization (29b) object-extraction occurs without a relative clause as compared to the baseline condition (29a), the canonical structure (Ben-Shachar et al. 2004). Here, again consistent activity was found in Broca's area, and has been interpreted as being due to higher computational costs involved in object-extracted structures.

(29) a. John natan ['et ha-sefer ha-'adom] [la-professor me-oxford]
John gave [the-book the-red] [to-the-professor from-Oxford]

b. ['et ha-sefer ha-'adom]i John natan ti [la-professor me-oxford]
[the-book the-red]i John gave ti [to-the-professor from-Oxford]

However, this BOLD effect in Broca's area was found only for a subset of (plausible) object-extracted clauses (Chen et al. 2006), in which the head (object) NP was animate and the subject NP of the relative clause was inanimate (30a), but not for the opposite order (30b).

(30) a. The director$_i$ that the movie pleased t_i received a prize at the film festival.

b. The movie$_i$ that the director watched t_i received a prize at the film festival.

These findings indicate that not necessarily higher computational costs involved in object-extracted clauses causing this effect but the revision of θ-role assignment.

The (re)assignment of an agent role to the object NP in the relative clause may conflict with a canonical heuristic, which expects an inanimate θ-role such as Theme. The effect found in Broca's area might be thus due to θ-role reassignment rather than to computations associated with object-extraction. However, we cannot exclude the interpretation that higher computational costs associated with this thematic revision causes the effect rather than the semantic process *per se*. While these neuroimaging studies further explore specific sentence structures that generate significant BOLD effects in Broca's area, methodological reasons cannot exclude the assumption of higher computational costs—although linked to specific sentence structures. Again, in Makuuchi et al.'s (2009) German study, double (31a) and single nested (32a) sentence structures were respectively compared against linear baseline sentences (31b, 32b) to manipulate syntactic complexity against distance (marked in bold).

(31) a. Peter wusste, dass Maria, [die Hans, der gut aussah, liebte], Johann geküßt hatte.
Peter knew that Maria, [who loved, who was good looking, Hans], kissed Johann.

b. Peter wusste, dass Achim den großen Mann gestern am späten Abend gesehen hatte.
Peter knew that Achim saw the tall man yesterday late at night.

(32) a. Peter wusste, dass Maria, [die weinte], Johann geküßt hatte und zwar gestern abend.
Peter knew that Maria, [who loved, who was good looking, Hans], kissed Johann.

b. Peter wusste, dass Achim den großen Mann gesehen hatte und zwar am Abend.
Peter knew that Achim saw the tall man at night and that was late.

The fMRI-DTI data support the view that syntax recruits the posterior parts of BA 44 while the inferior frontal sulcus located above Broca's area is involved in memory processes. To avoid those and other methodological issues, researchers examined structural properties of an artificial language to be learned (Musso et al. 2003; Tettamanti et al. 2009). In Musso and colleagues' study, BA 45 was activated when subjects performed on a possible syntactic rule found in natural languages (but not in the subjects' native language) when compared with syntactic patterns not found in any natural languages. However, precise conclusions cannot be drawn from these results as further studies are required to verify that BA 45 has a specific role in processing novel syntax or structures as compared to BA 44 or other relevant areas of interest.

Again, Tettmanti et al.'s study contrasted learning of linear (rigid) and hierarchical (non-rigid) structures in a visuo-spatial task, which consisted of linear sequences of non-symbolic elements. A domain-independent effect was found for the acquisition of non-rigid syntax in the left BA 44, but not for rigid syntax (see also Fiebach and Schubotz 2006). Accordingly, it has been suggested that hierarchical structures are specifically recruited in Broca's area and other syntactic operations outside of Broca's area: local phrase structure building (frontal operculum and anterior superior temporal gyrus, STG) and syntactic integration takes place in the posterior STG/STS (Anwander et al. 2007; see also Fig. 9.1).

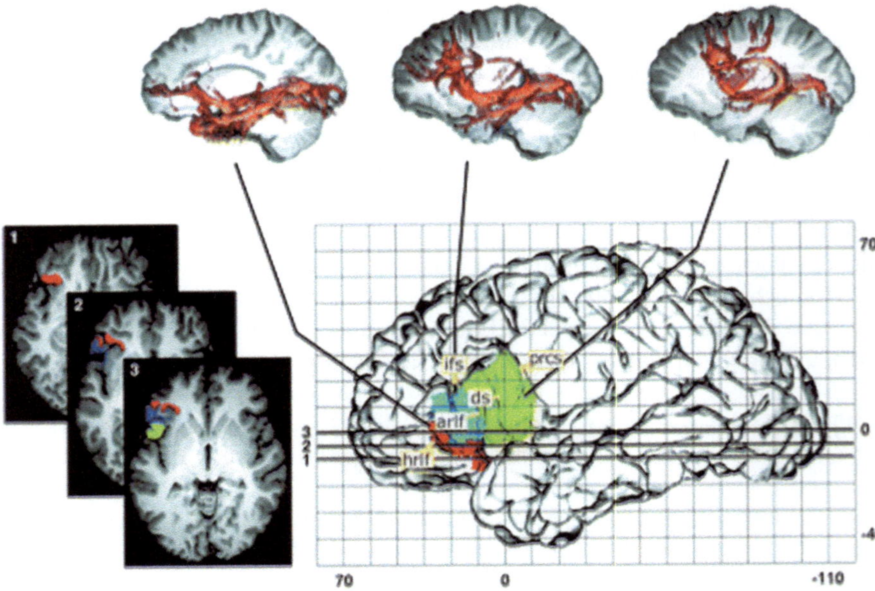

Fig. 9.1 Single subject segmentation of the left IFG into fronto operculum (red), BA 45 (blue) and BA 44 (green) and connectivity projections from these areas to the temporo-parietal cortex. Arlf ascending branch of the lateral fissure, ds diagonal sulcus, horizontal branch of the lateral fissure hrlf, inferior frontal sulcus ifs, precentral sulcus prcs. (Adapted, Anwander et al. 2007; © Oxford University Press)

The fronto-temporal network is also involved in disambiguating sentence structures (Tyler et al. 2011).

(33) a. The newspaper reported that [bullying teenagers] are a problem for the local school.
b. The newspaper reported that bullying [teenagers] is bad for their self-esteem.

Study participants passively listened to sentences, which were disambiguated at the verb of the (right-branching) relative clause as agreement permitted only one reading. Prior to this verb, an *ing*-word was presented together with a noun (e.g., *bulling teenagers*), which can function as an adjective (33a) or as a gerund (33b). The authors report left-sided co-activation of BAs 45 and 47 and the posterior MTG, and attributed a left parietal cluster to WM demands. Although disambiguation per se is a pure syntactic process, semantic factors such as initial preference strategies may have contributed to the comprehension process. However, Tyler and colleagues also emphasize that the findings seem to be supported by P600 peaks found in EEG/MEG[4]

[4] The most popular non-invasive method to measure electrophysiological activity of the brain is called event-related potentials (ERPs). It can be considered as functional electroencephalography (EEG) as electric cortical activity is measured in response to a cognitive-behavioral task, whereas electrodes are placed on the scalps surface. The EEG was discovered by the German physician Hans Berger in 1924. It reflects thousands of parallel cortical processes and correlation of the

studies on syntactic ambiguity.[5] Thus, the P600 might reflect the interaction between the left inferior frontal gyrus (IFG) and the middle temporal gyrus (MTG).

The brief overview shows that there is a wide range of different linguistic and cognitive hypotheses about how the brain computes certain aspects of syntax. However, syntactic computations take mainly place in the left fronto-temporal net and this net is used for different types of computations involving domain-independent syntactic processes, interfaces to non-syntactic computations and multiple factors that contribute to sentence processing. In particular, the left pars operculum seems to be the site for manipulating short-term dependency relations. Most challenging is to disentangle differevnt dimensions involved in sentence processing including task-specific requirements, memory load, semantics (or other context information), access and integration of morphological and syntactic substructures.

As discussed before, mainly the left-sided superior and medial temporal lobe (S/MTL) seem to be recruited during sentential syntactic integration and ambiguity resolution. Both aspects involve semantic processes, which indicate that temporal regions elicit activations at the interface of syntactic and semantic processes. Price's (2010) review of 100 fMRI studies published in 2009 shows that four key regions were reported with respect to grammatically correct sentences being either plausible or not: left MTG, bilateral anterior temporal pole, left angular gyrus (AG) and the posterior cingulate/precuneus (see also Binder et al. 1997).

Without addressing the single experimental MRI conditions of each study, the general picture provided by Price shows that comprehension difficulties closely correlate with computational costs occurs. Thus, in the case of computing semantic

---

electric signal to a specific stimulus requires many trials that random noise can be averaged out. The ERPs provide an online measurement of the brain's activity and may reveal responses, which cannot be exclusively detected by behavioral means. While the temporal resolution of ERPs is excellent (ca. <10 ms), their spatial resolution remains undefined as ERPs cannot be (sub)cortically localized. The most known ERP components are the early left anterior negativity (ELAN), the N400 and the P600. ELAN is a negative μV response that peaks ca. <200 ms after presentation of a phrase structure violation (e.g., *Sam played on the *wrote*) and the N400 is a negative response to a semantic violation ca. 400 ms after the onset of the stimulus presentation (e.g., **Sam ate the shoes*); the P600 is a positive response (also called syntactic positive shift, SPS) that will be elicited at around 500 ms after stimulus presentation and peaks at ca. 600 ms; it can be measured in garden-path sentences, at gap-filling dependencies and when morpho-syntactic violations (e.g., number, case, gender) are encountered (Osterhout and Holcomb 1992).

[5] Magnetoencephalography (MEG), first reported by Cohen (1968), has a temporal resolution and generates evoked responses much like EEG/ERP. The magnetic components are labeled according to temporal latency. For example, the M100 is elicited ca. 100 ms post-stimulus presentation of a particular stimulus, usually tones, phonology information or words. The M400 (corresponds to the N400) is generated in context of semantic processing. However, magnetic fields are less distorted than EEG and have therefore a better spatial resolution. While EEG is sensitive to extracellular volume currents elicited by post-synaptic potentials, MEG is sensitive to intracellular currents of these synaptic potentials. EEG can detect activity in the sulci and at top of the cortical gyri, but MEG detects activity mostly in the sulci. In contrast to EEG, MEG activity can be localized with more accuracy. MEG is often combined with (f)MRI to generate functional cortical maps.

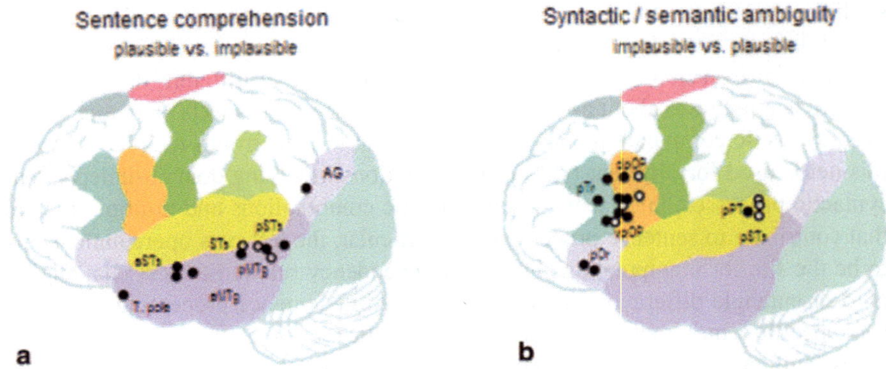

Fig. 9.2 Sentences comprehension and ambiguities: summary of left-sided activation foci found in numerous fMRI studies **a** dark circles, plausible vs. implausible sentences; white circles, sentences with vs. without visual gestures. A anterior, p posterior, STs superior temporal sulcus, MTg middle temporal gyrus, T. pole temporal pole, AG angular gyrus **b** dark spots, semantic ambiguity; white spots, syntactic ambiguity. PTr pars triangularis, d dorsal, v ventral, pOP pars opercularis, pOr par orbitalis, PT planum temporale. (Adapted and modified, Price 2010; © 2010 New York Academy of Sciences)

(Obleser and Kotz 2010; Bilenko et al. 2009) or syntactic ambiguities (e.g., Richardson et al. 2010; Makuuchi et al. 2007) or if implausible vs. plausible meanings are compared, activities were particularly reported in the left pars opercularis (BA 44). However, when comparing plausible vs. implausible meanings (e.g., Rogalsky and Hickok 2009; Devauchelle et al. 2009), activations take mostly place in the superior or middle temporal region (Fig. 9.2).

The question arises whether we can find specific neural circuits supporting the computation of semantic structures in terms of verb argument structures, which represent the interface between syntax and semantics at the sentence level. Most neuroimaging studies report the involvement of the left-sided MTG and Broca's area in computing verb V-argument structures (e.g., Fiez 1997; Thompson-Schill et al. 1997). Again, typically an increase of sentence complexity is reflected in recruiting Broca's areas. The significant contribution of linguistic context to verb processing was demonstrated by an MEG study (Assadollahi and Rockstroh 2008). In this study, processing single verbs generate in the range of 250–300 ms different activation levels in the MTG depending on the argument structure associated with a verb. However, when the verbs were presented together with a proper noun as grammatical subject, the activation shifted to 350–440 ms and additional activation was reported for Broca's area.[6] Interestingly, intransitive verbs (*to snore*) elicited higher activity than ditransitive verbs (*to give*) in the left MTG and Broca's area. The authors argue that the variance in activity may be due to a completeness effect: The left IFG may integrate semantic-syntactic information,

[6] MEG or EEG data can be mapped onto a standard anatomical brain (inverse mapping).

which is complete in the case of intransitive verbs. However, for transitive and ditransitive verbs object arguments cannot be assigned and thus θ-roles will not be mapped to phrase structures or in other words, semantic integration does not take place. However, others have argued that the integration of semantic information is mainly a domain of the temporo-parietal region (e.g., Thompson et al. 2007).

However, in considering the various electrophysiological and neuroimaging results mentioned above, the fronto-temporal network seems to support semantic-syntactic computations. The MTG generates θ-roles, which will be mapped to syntactic structures in the left IFG, whereas BA 47 might specifically contribute to this process. The integration of novel, atypical or even implausible meanings, the process of resolving semantic and/or syntactic ambiguities and other semantic computations, that require WM resources, may require looping within the fronto-temporal network. The neuroimaging data show a sketchy picture but reveal that the left IFG can be subdivided into semantic and syntactic functions. While parsing in general and specifically with respect to the generation of hierarchical structures is a left-sided domain of the pars opercularis (BA 44), semantic integration in terms of mapping to syntactic information takes place in Broca's area and in the orbital area (BA 47), located below pars triangularis, and in the frontopolar area (BA 10)—the anterior portion of the prefrontal cortex (e.g., Démonet et al. 1992; Demb et al. 1995; Gabrieli et al. 1996; Binder et al. 1997; Newman et al. 2003).

Further research is required to specify the neural correlates of particular semantic and syntactic structures at the sentence level. Based on a multi-receptor approach, Amunts et al. (2010) offered a new subdivision of Broca's area, but it remains to be seen whether this neural segmentation can be mapped onto functional computations. At present, we can only speculate how different semantic-syntactic functions are scaffolded by different neural areas, how these areas interact within the sentence processing net and which aspects should be considered as domain-specific or which aspects are shared with non-linguistic computations of other cognitive domains. Finally, the empirically findings might support a theoretical approach, which does not consider separate semantic computations at the sentence level, but supports the view of computations directly mapping conceptual and syntactic representations.

9.2 The Role of Working Memory

Different WM demands vary with the structural complexity of a sentence. Often differences in sentence complexity correlate directly with different WM demands. To provide an example: Typically, object relative clause (object-RC) sentences are structurally more complex than subject-RC sentences. As shown in (33) both sentence types differ in syntactic complexity (IP, inflectional phrase; CP, constitutional phrase; Chomsky 1986).

(34) a. [IPThe researcheri [CP thati [IP _ i attacked the senator]] admitted the error]
b. [IPThe senatori [CP thati [IP the researcher attacked _ i]] admitted the error]

In subject RC sentences (34a), the agent of the matrix clause is also the agent of the relative clause. Thus, the canonical subject-verb object structure (in English) has been maintained, since the subject of the matrix clause shifted to the relative clause. However, comprehending an object RC sentence (34b) requires more "cognitive costs" than a subject RC sentence, because the structure includes an empty syntactic category (trace) in object position of the RC (see Gibson 1998). It is the agent of the matrix clause that functions as object of the relative clause. Thus, subjects' listing span should be reduced for object RC sentence compared to subject RC sentences. Another example are ambiguous syntactic structure, in which the relative clause can be attached high to the first NP or low to the second NP, the more recent phrase (35).

(35) [The daughter$_{i\,(NP1)}$ of the singer$_{j\,(NP2)}$ [who$_{i/j\,(NP1/2)}$ [$_{IP}$_$_{i/j}$ met the teacher]] found the key]

Attachment preferences seem to vary across languages. For instance, an English speaker prefers to attach the RC to the lower NP2, but a Dutch, German, French, or Spanish speaker tends to prefer the high NP1 (e.g., Cuetos and Mitchell 1988; Brysbaert and Mitchell 1996; Zagar et al. 1997). Also, it has been reported that children with low WM span were indeed more likely to select the lower NP than were children with a high WM span (Felser et al. 2003). This recency effect was originally observed by Kimball (1973) and revised by Frazier (1979) in the garden-path model (late-closure). Thus, we cannot exclude the scenario that attachment preferences are determined by individual WM span. Subjects with a low WM span may prefer late closure while subjects with a high WM span may tend to attach the RC to the initial NP.

Miller and Chomsky (1963) were among the first who noted that object RC sentences (36a) require significantly more processing costs than subject RC sentences (36b).

(36) a. The fireman who the cop speedily rescued sued the city over working conditions.
b. The fireman who speedily rescued the cop sued the city over working conditions.

According to Chomsky's Government Binding theory (1981) object and subject RC sentences can be described as follows by using X' notation (see 2 and 3; constituent phrase (CP), inflectional phrase (IP), noun phrase (NP); verb phrase (VP); however, for illustrative reasons the figures do not show intermediate nodes of the X-bar tree structure). The term Subject in subject RC refers to the fact that the Subject of the matrix clause functions as Subject in the RC (37). The term Object in object RC refers to the fact that the Subject of the matrix clause functions as Object in the relative clause. The object NP of the RC is not lexicalized, that is, it is an empty syntactic category (*t*). The relative pronoun moves to the first position of the RC and leaves behind a trace in object position of the RC.

(37)

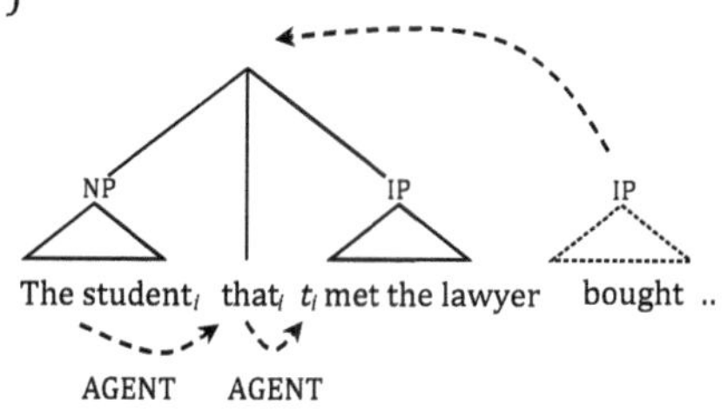

Because the subject NP of the matrix clause is co-indexed with the relative pronoun, the listener re-activates at trace position the co-indexed NP. The increase of processing costs in object RC sentences compared to subject RC sentences has been reported in several studies (e.g., Holmes and O'Regan 1981; Ford 1983; King and Just 1991). In subject RC sentences the relative pronoun moves one level higher out of the subject position of the RC.

However, the subject NP of the RC is filled by the preceding relative pronoun, which in turn is co-indexed with the preceding subject NP of the matrix clause. According to one process account, the time required to maintain in WM the subject NP of the matrix clause (before re-activation occurs) is much longer for object RC sentences than for subject RC sentences (Wanner and Maratsos 1978). In particular, the increase of cognitive costs in object RC sentences may occur at the immediate vicinity of the trace position, that is, between the verb of the RC and the verb of the matrix clause. The listener needs to re-activate at the trace position the antecedent and must assign a different thematic role to comprehend the sentence (38).

(38)

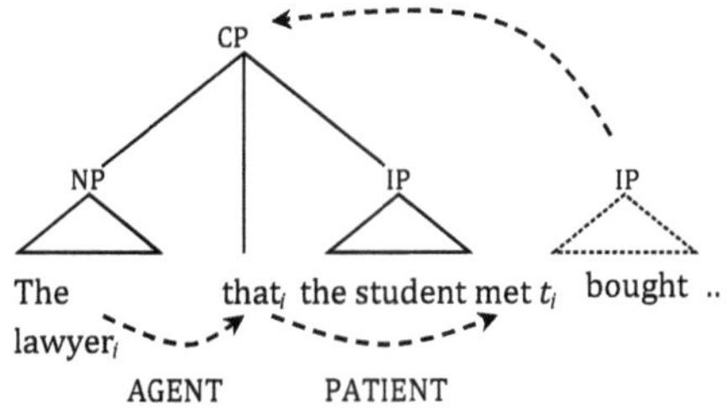

In using the cross-modal lexical decision task[7], spoken comprehension of object RC sentences in college-aged speakers revealed re-activation (priming) of the antecedent (*boy*) at the trace position (t_i), but no priming of the Subject of the relative clause (*crowd*). (e.g., Nicol and Swinney 1989):

(39) The policeman saw the boy$_i$ that$_i$ the crowd at the party accused t_i of the crime.

These responses do not reflect a residual effect but re-activation of the object is supported by a non-priming effect of the Object at pre-trace position (immediately before the verb of the RC). However, thematic role assignment can begin earlier before the trace position (Weckerly and Kutas 1999).

Language perceivers with a lower WM span prefer to connect incoming words to the most recently processed phrase (Kimball 1973; Frazier 1979; Gibson 1998). Sentence (40) is difficult to understand because of late closure. The adverb *yesterday* must be attached to the VP of the non-recent matrix verb *said*, but not to the recent VP *will leave for good* to understand the sentence.

(40) Marcel said that Susan will leave for good yesterday.

People with a high WM span have less difficulty with those sentence structures, because they are able to hold more words in their buffer. The recency principle or minimal attachment (late closure) may rely on limited WM resources available for sentence processing. If perceived words are connected to the most recent processes constituent, the risk of losing information due to inferences or decay is minimized (e.g., Frazier 1979, 1987; Frazier and Fodor 1978). That the recency effect is not universal, but a language-specific factor was demonstrated by Cuetos and Mitchell (1988). They found in contrast to speakers of English that speakers of Spanish had a small but significant preference to attach the more distant NP rather than the most recent one (see also Brysbaert and Mitchell 1996; Desmet et al. 2002). Frazier and Clifton (1997) proposed therefore the *construal model* of language processing that considers not only grammatical constrains but also semantic information associated with thematic roles. So far, it has been rarely tested that individuals with limited resources rely more or heavily on late closure. However, in one study it has been reported that 6–7-year old children with a lower WM span tend indeed to rely on a recency strategy by attaching the RC to the most recent NP (Felser et al. 2003).

[7] The cross-modal lexical decision (CMLD) task is considered as an online method to examine early access to lexical (or other linguistic) information. It is a dual task because participants typically listening to individual sentences while performing simultaneously a lexical decision task (the decision whether a lexical string is a word or not) at a certain point during the sentence presentation. If in the lexical decision task a word is displayed, it can be semantically related to a word or structure in the sentence or it is unrelated. The statistical evaluation of the reaction time differences between related and control words show whether lexical priming was found or not (Swinney and Hakes 1976).

Thus, in (40) children with a lower WM span prefer to attach the preceding NP2 *actress* to the RC rather than the NP1 *servant.*

(41) Someone shot the servant of the actress who was on the balcony

Independent of a semantic plausibility effect, late closure may be indeed related to a reduced WM span (42).

(42)

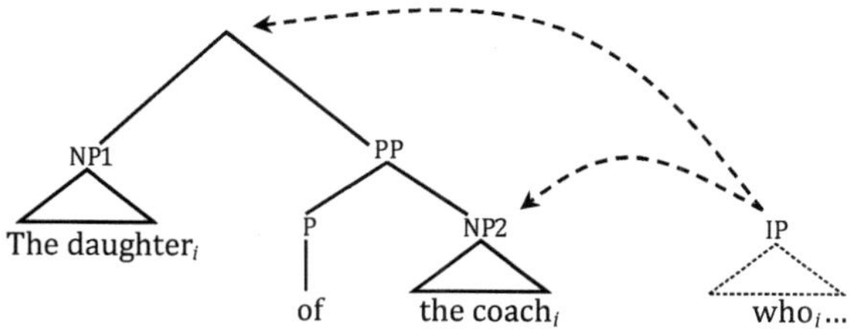

It might be related to compensatory mechanisms of limited WM resources, if study participants rely completely on a recency strategy (lower NP). De facto, as WM spans vary between participants and across the life-cycle, the availability of sentence processing capacities depends on a range of different subject-specific variables.

Let us turn to the neural correlates of WM functions. In one fMRI study object-embedded sentence structures selectively activated in contrast to subject-embedded sentence structures three frontal areas: left IFG (Broca's area), left dorsal premotor cortex and left supplementary motor area, both BA 6 (Meltzer et al. 2010). Both BA 6 areas were only active in reversible sentence. The reason may be that parsing semantically reversible object-embedded sentences is associated with an increased cognitive demand. The involvement of both the premotor and supplementary motor area may reflect computations, which operate in a more controlled fashion. In contrast, Broca's area is particularly sensitive to syntactic complexity such as syntactic movements or other structures deviating from the "animate-agent-first-word order." Moreover, the analysis of reversible vs. irreversible sentence structures shows that in addition to the left prefrontal structures temporal and parietal regions were engaged. This effect points to increased attentional computations during θ-role assignment (Fig. 9.3).

Neuroimaging studies suggest that verbal WM functions are supported by different neural areas (e.g., Smith and Jonides 1998; Capeza and Nyberg 2000; see for a review Chein et al. 2003): the left (sometimes in addition the right) IFG (Broca's area: BAs 44 and 45), the left right middle frontal gyrus (BAs 9 and 46), the left and right medial frontal cortex, in particular the premotor cortex and the pre-supplementary motor area (BA 6), the bilateral supramarginal gyrus (SMG) in the inferior parietal lobe (BA 40), the anterior cingulate cortex (ACC) and the cerebrellum.

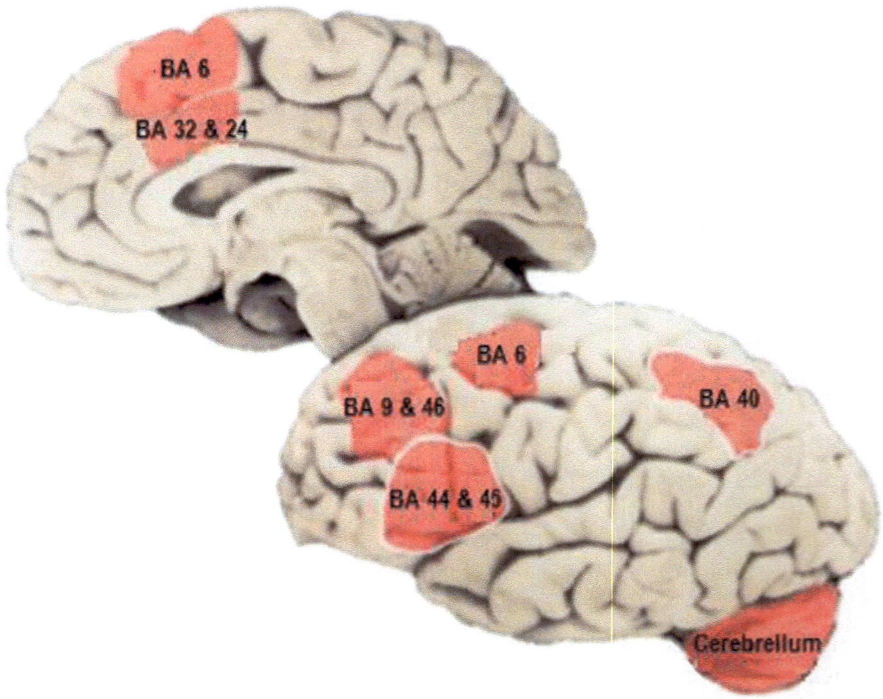

Fig. 9.3 Mid-sagittal (above) and sagittal view (below) of Brodmann areas and cerebrellum related to verbal working memory. (Adapted and modified, Chein et al. 2003; © Elsevier Limited)

However, it has been emphasized that the right prefrontal cortex is involved in spatial and non spatial WM, but the left prefrontal cortex involves only verbal WM. The involvement of temporal-parietal regions seems to depend on the specific stimuli to be computed (Prabhakaran et al. 2000).

The consistent activation of Broca's area in verbal WM tasks speaks in favor of the assumption that the IFG function is *also* related to speech-related rehearsal operations. The involvement of Broca's area during complex sentence processing such as reanalysis is apparent as we mentioned above. Moreover, some studies indicate that pars opercularis is specialized for rehearsing non-canonical syntactic structures, and pars orbitalis for rehearsing lexical semantic information (Just et al. 1996; Dapretto and Bookheimer 1999). In contrast, the parietal cortex may be activated in WM tasks requiring attention including the inhibition of automatically computed verbal information (e.g., Posner and Petersen 1990; Meyer et al. 2013).

The examination of sentence processing involves multiple variables, which are at the empirical level closely conjoined or refer to a single phenomenon. For instance, the computation of syntactic structures along with θ-roles assigned to the NPs requires to some extent WM functions. Structure (syntax) and content (words) merge to generate a plausible string of units. The cognitive mechanisms, which enable these computations, may not exist without these computations. It remains to be seen whether the neural substrates scaffolding the process can be subdivided along the line of these different variables or whether future groundwork can provide evidence for a more specific neuronal net model of sentence processing.

References

Amunts, K., Lenzen, M., Friederici, A. D., Schleicher, A., Morosan, P., Palomero-Gallagher, N., & Zilles, K. (2010). Broca's region: Novel organizational principles and multiple receptor mapping. *PLoS Biology, 8*(9). doi:10.1371/journal.pbio.1000489.

Anwander, A., Tittgemeyer, M., von Cramon, D. Y., Friederici, A. D., & Knösche, T. R. (2007). Connectivity-based parcellation of Broca's yarea. *Cerebral Cortex, 17*(4), 816–825.

Assadollahi, R., & Rockstroh, B. S. (2008). Representation of the verb's argument-structure in the human brain. *BMC Neuroscience, 9*, 69.

Ben-Shachar, M., Palti, D., & Grodzinsky, Y. (2004). Neural correlates of syntactic movement: Converging evidence from two fMRI experiments. *NeuroImage, 21*(4), 1320–1336.

Bilenko, N. Y., Grindrod, C. M., Myers, E. B., & Blumstein, S. E. (2009). Neural correlates of semantic competition during processing of ambiguous words. *Journal of Cognitive Neuroscience, 21*(5), 960–975.

Binder, J. R., Frost, J. A., Hammeke, T. A., Cox, R. W., Rao, S. M., & Prieto, T. (1997). Human brain language areas identified by functional magnetic resonance imaging. *The Journal of Neuroscience, 17*(1), 353–362.

Bookheimer, S. (2002). Functional MRI of language: New approaches to understanding the cortical organization of semantic processing. *Annual Review Neuroscience, 25*, 151–188.

Brysbaert, M., & Mitchell, D. C. (1996). Modifier attachment in sentence processing: Evidence from Dutch. *Quarterly Journal of Experimental Psychology, 49A*, 664–669.

Burton, M. (2001). The role of inferior frontal cortex in phonological processing. *Cognitive Science, 25*, 695–709.

Capeza, R., & Nyberg, L. (2000). Neural bases of learning and memory: Functional neuroimaging evidence. *Current Opinion in Neurology, 4*, 415–421.

Caplan, D., Alpert, N., & Waters, G. (1998). Effects of syntactic structure and propositional number on patterns of regional cerebral blood flow. *Journal of Cognitive Neuroscience, 10*(4), 541–552.

Caplan, D., Chen, E., & Waters, G. (2008). Task-dependent and task-independent neurovascular responses to syntactic processing. *Cortex: A journal devoted to the study of the nervous system and behavior, 44*(3), 257–275.

Chein, J. M., Ravizza, S. M., & Fiez, J. A. (2003). Using neuroimaging to evaluate models of working memory and their implications for language processing. *Journal of Neurolinguistics, 16*(4-5), 315–339.

Chen, E., West, W. C., Waters, G., & Caplan, D. (2006). Determinants of bold signal correlates of processing object-extracted relative clauses. *Cortex: A journal devoted to the study of the nervous system and behavior, 42*(4), 591–604.

Chomsky, N. (1981). *Lectures on government and binding*. Dordrecht: Foris Publications.

Chomsky, N. (1986). *Knowledge of language: Its nature, origin, and use*. New York: Praeger.

Cohen D. (1968). Magnetoencephalography: Evidence of magnetic fields produced by alpha rhythm currents. *Science*, 161, 784–786.

Cuetos, F., & Mitchell, D. C. (1988). Cross-linguistic differences in parsing: Restrictions on the use of the Late Closure strategy in Spanish. *Cognition, 30*(1), 73–105.

Dapretto, M., & Bookheimer, S. Y. (1999). Form and content: Dissociating syntax and semantics in sentence comprehension. *Neuron, 24*, 427–432.

Demb, J. B., Desmond, J. E., Wagner, A. D., Vaidya, C. J., Glover, G. H., & Gabrieli, J. D. (1995). Semantic encoding and retrieval in the left inferior prefrontal cortex: A functional MRI study of task difficulty and process specificity. *The Journal of Neuroscience, 15*(9), 5870–5878.

Démonet, J.-F., Chollet, F., Ramsay, S., Cardebat, D., Nespoulous, J.-L., Wise, R., Rascol, A., & Frackowiak, R. (1992). The anatomy of phonological and semantic processing in normal subjects. *Brain: A journal of neurology, 115*(6), 1753–1768.

Desmet, T., De, B. C., & Brysbaert, M. (2002). The influence of referential discourse context on modifier attachment in Dutch. *Memory & Cognition, 30*(1), 150–157.

Devauchelle, A.-D., Oppenheim, C., Rizzi, L., Dehaene, S., & Pallier, C. (2009). Sentence syntax and content in the human temporal lobe: An fMRI adaptation study in auditory and visual modalities. *Journal of Cognitive Neuroscience, 21*(5), 1000–1012.

Dronkers, N. F., Wilkins, D. P., Redfem, B. B., Van Valin, J. R., & Jaeger, J. J. (1994). A reconsideration of the brain areas involved in the disruption of morphosyntactic comprehension. *Brain and Language, 47,* 461–463.

Embick, D., Marantz, A., Miyashita, Y., O'Neil, W., & Sakai, K. L. (2000). A syntactic specialization in Broca's area. *Proceedings of the National Academy of Science, 97*(11), 65150–65154.

Felser, C., Marinis, T., & Clahsen, H. (2003). Children's processing of ambiguous sentences: A study of relative clause attachment. *Language Acquisition, 11*(3), 127–163.

Fiebach, C. J., & Schubotz, R. I. (2006). Dynamic anticipatory processing of hierarchical sequential events: A common role for Broca's area and ventral premotor cortex across domains? *Cortex: A journal devoted to the study of the nervous system and behavior, 42*(4), 499–502.

Fiez, J. A. (1997). Phonology, semantics, and the role of the left inferior prefrontal cortex. *Human Brain Mapping, 5*(2), 79–83.

Ford, M. (1983). A method for obtaining measures of local parsing complexity throughout sentences. *Journal of Verbal Learning and Verbal Behavior, 22*(2), 203–218.

Frazier, L. (1979). *On comprehending sentences: Syntactic parsing strategies*. Bloomington: Indiana University Linguistics Club.

Frazier, L. (1987). Structure in auditory word recognition. *Cognition, 25*(1-2), 416–422.

Frazier, L., & Clifton, C. (1997). Construal: Overview, motivation, and some new evidence. *Journal of Psycholinguistic Research, 26*(3), 277–295.

Frazier, L., & Fodor, J. D. (1978). The sausage machine: A new two-stage parsing model. *Cognition, 6,* 291–325.

Gabrieli, J. D. E., Desmond, J. E., Demb, J. B., Wagner, A. D., Stone, M. V., Vaidya, C. J., & Glover, G. H. (1996). Functional magnetic resonance imaging of semantic memory processes in the frontal lobes. *Psychological Science, 7*(5), 278–283.

Gibson, E. (1998). Syntactic complexity: Locality of syntactic dependencies. *Cognition, 68*(1), 1–76.

Goodglass, H. (1993). *Understanding aphasia*. San Diego: Academic Press.

Herdener, M., Humbel, T., Esposito, F., Habermeyer, B., Cattapan-Ludewig, K., & Seifritz, E. (2012). Jazz drummers recruit language-specific areas for the processing of rhythmic structure. *Cerebral Cortex, 24*(3):836–843. (ePrint Nov. 25).

Hillert, D. (1990). *Sprachprozesse und Wissensstrukturen: Neuropsychologische Grundlagen der Kognition* [German]. Opladen: Westdeutscher Verlag.

Hillert, D. (Ed.) (1994). *Linguistics and cognitive neuroscience. Linguistische Berichte, Special Issue 4*. Opladen: Westdeutscher Verlag.

Hillert, D., & Buračas, G. (2009). The neural substrates of spoken idiom comprehension. *Language and Cognitive Processes, 24*(9), 1370–1391.

Holmes, V. N., & O'Regan, J. K. (1981). Eye fixation patterns during the reading of relative clause sentences. *Journal of Verbal Learning and Verbal Behavior, 20,* 417–430.

Iacoboni, M., Woods, R. P., Brass, M., Bekkering, H., Mazziotta, J. C., & Rizzolatti, G. (1999). Cortical mechanisms of human imitation. *Science, 286,* 2526–2528.

Just, M. A., Carpenter, P. A., Keller, T. A., Eddy, W. F., & Thulborn, K. R. (1996). Brain activation modulated by sentence comprehension. *Science, 274*(5284), 114–116.

Kang, A. M., Constable, R. T., Gore, J. C., & Avrutin, S. (1999). An event-related fMRI study of implicit phrase-level syntactic and semantic processing. *NeuroImage, 10*(5), 555–561.

Kimball, J. (1973). Seven principles of surface structure parsing in natural language. *Cognition, 2,* 15–47.

King, J., & Just, M. A. (1991). Individual differences in syntactic processing: The role of working memory. *Journal of Memory and Language, 30,* 580–602.

Makuuchi, M., Bahlmann, J., Anwander, A., & Friederici, A. D. (2007). Segregating the core computational faculty of human language from working memory. *Proceedings of the National Academy of Science, 106*(20), 8362–8367.

Mazoyer, B. M., Tzourio, N., Frak, V., Syrota, A., Murayama, N., Levrier, O., Salamon, G., Dehaene, S., Cohen, L., & Mehler, J. (1993). The cortical representation of speech. *Journal of Cognitive Neuroscience, 5*(4), 467–479.

Meltzer, J. A., McArdle, J. J., Schafer, R. J., & Braun, A. R. (2010). Neural Aspects of Sentence Comprehension: Syntactic Complexity, Reversibility, and Reanalysis. *Cerebral Cortex, 20*(8), 1853–1864. doi:10.1093/cercor/bhp249.

Meyer, M., Friederici, A. D., & von Cramon, D. Y. (2000). Neurocognition of auditory sentence comprehension: Event related fMRI reveals sensitivity to syntactic violations and task demands. *Brain Research. Cognitive Brain Research, 9*(1), 19–33.

Meyer, L., Obleser, J., & Friederici, A. D. (2013). Left parietal alpha enhancement during working memory-intensive sentence processing. *Cortex: A journal devoted to the study of the nervous system and behavior, 49*(3), 711–721.

Miller, G. A., & Chomsky, N. (1963). Finitary models of language users'. In (Eds.), Luce, R. D., Bush, R. R. & E. Galanter, *Handbook of Mathematical Psychology,* (vol. II). New York, NY: Wiley: 419–491.

Moro, A., Tettamanti, M., Perani, D., Donati, C., Cappa, S. F., & Fazio, F. (2001). Syntax and the brain: Disentangling grammar by selective anomalies. *NeuroImage, 13*(1), 110–118.

Musso, M., Moro, A., Glauche, V., Rijntjes, M., Reichenbach, J., Buechel, C., & Weiller, C. (2003). Broca's area and the language instinct. *Nature Neuroscience, 6,* 774–781.

Newman, S. D., Just, M. A., Keller, T. A., Roth, J., & Carpenter, P. A. (2003). Differential effects of syntactic and semantic processing on the subregions of Broca's area. *Cognitive Brain Research, 16*(2), 297–307.

Nicol, J., & Swinney, D. (1989). The role of structure in coreference assignment during sentence comprehension. *Journal of Psycholinguistic Research, 18*(1), 5–19.

Obleser, J., & Kotz, S. A. (2010). Expectancy constraints in degraded speech modulate the language comprehension network. *Cerebral Cortex, 20*(3), 633–640.

Posner, M. I., & Petersen, S. E. (1990). The attention system of the human brain. *Annual Review of Neuroscience, 13*(1), 25–42.

Prabhakaran, V., Narayanan, K., Zhao, Z., & Gabrieli, J. D. E. (2000). Integration of diverse information in working memory within the frontal lobe. *Nature Neuroscience, 3*(1), 85–90.

Price, C. J. (2010). The anatomy of language: A review of 100 fMRI studies published in 2009. *Annals of the New York Academy of Sciences, 1191,* 62–88.

Richardson, F. M., Thomas, M. S. C., & Price, C. J. (2010). Neuronal activation for semantically reversible sentences. *Journal of Cognitive Neuroscience, 22*(6), 1283–1298.

Rogalsky, C., & Hickok, G. (2009). Selective attention to semantic and syntactic features modulates sentence processing networks in anterior temporal cortex. *Cerebral Cortex, 19*(4), 786–796.

Smith, E. E., & Jonides, J. (1998). Neuroimaging analyses of human working memory. *Proceedings of the National Academy of Sciences of the USA, 95*(20), 12061–12068.

Stromswold, K., Caplan, D., Alpert, N., & Rauch, S. (1996). Localization of syntactic comprehension by positron emission tomography. *Brain and Language, 52*(3), 452–473.

Swinney, D. A., & Hakes, D. T. (1976). Effects of prior context upon lexical access during sentence comprehension. *Journal of Verbal Learning and Verbal Behavior, 15,* 681–689.

Tettamanti, M., Rotondi, I., Perani, D., Scotti, G., Fazio, F., Cappa, S. F., & Moro, A. (2009). Syntax without language: Neurobiological evidence for cross-domain syntactic computations. *Cortex: A journal devoted to the study of the nervous system and behavior, 45*(7), 825–838.

Thompson, C. K., Bonakdarpour, B., Fix, S. C., Blumenfeld, H. K., Parrish, T. B., Gitelman, D. R., & Mesulam, M.-M. (2007). Neural correlates of verb argument structure processing. *Journal of Cognitive Neuroscience, 19*(11), 1753–1767.

Thompson-Schill, S. L., D'Esposito, M., Aguirre, G. K., & Farah, M. J. (1997). Role of left inferior prefrontal cortex in retrieval of semantic knowledge: A reevaluation. *Proceedings of the National Academy of Sciences, 94*(26), 14792–14797.

Tyler, L. K., Marslen-Wilson, W. D., Randall, B., Wright, P., Devereux, B. J., Zhuang, J., Papoutsi, M., & Stamatakis, E. A. (2011). Left inferior frontal cortex and syntax: Function, structure and behaviour in patients with left hemisphere damage. *Brain: A Journal of Neurology, 134*(2), 415–431.

Vul, E., Harris, C., Winkielman, P., & Pashler, H. (2009). Puzzlingly high correlations in fMRI studies of emotion, personality, and social sognition. *Perspectives on Psychological Science, 4*(3), 274–290.

Wanner, E., & Maratsos, M. (1978). An ATN approach in comprehension. In M. Halle, J. Bresnan, & G. Miller (Eds.), *Linguistic theory and psychological reality* (pp. 119–161). Cambridge: MIT Press.

Weckerly, J., & Kutas, M. (1999). An electrophysiological analysis of animacy effects in the processing of object relative sentences. *Psychophysiology, 36*(5), 559–570.

Zagar, D., Pynte, J., & Rativeau, S. (1997). Evidence for early closure attachment on first-pass reading times in French. *Quarterly Journal of Experimental Psychology, 50A,* 421–438.

Chapter 10
Accessing Word Meanings

10.1 Lexical Concepts

The question whether the cortical regions recruiting noun and verb processing are segregated is central to the discussion of lexical computations in the brain. Single case lesion studies seem to indicate that in naming, concrete nouns might be processed in left anterior and middle temporal lobe and concrete (action) verbs in the premotor cortex (Damasio and Tranel 1993; see also McCarthy and Warrington 1985; Daniele et al. 1994). Other lesion studies do not speak in favor of a neural segregation for concrete verbs and nouns (e.g., De Renzi and Pellegrino 1995; Aggujaro et al. 2006).

Similarly, based on category- and/or modality-specific impairments, mostly found in semantic dementia, numerous single case lesion studies argue that the lexical semantic system is fractionated (e.g., Beauvois et al. 1978; Beauvois 1982; Warrington and Shallice 1984; McCarthy and Warrington 1988; Farah et al. 1989; Hart and Gordon 1992; Goodglass 1994; McCarthy and Warrington 1994; Grossman et al. 2013; Libon et al. 2013). However, the interpretation of these apparently selective impairments is less clear. Multiple factors can contribute to processes, which are selectively spared or impaired due to specific neurological events in an individual patient (e.g., Hillert 1990, 1992; Marques et al. 2013; Crepaldi et al. 2013). Here are a few examples:

In Beauvois et al's (1978) classical study, the object naming performance of the patient RG is described in the tactile, visual, and auditory modality. RG had an error rate of 29 % (71/100) in the right-handed tactile object naming task and that of 36 % (64/100) in the left-handed tactile object naming task. In contrast, he showed relatively intact naming in the visual (96/100) and auditory modality (79/100). Moreover, despite of this bilateral tactile anomia, RG was able to indicate the use of the unnamed object. However, this does not mean that RG was able to recognize the object as the meaning associated with the pantomime can be exclusively perceptually based. In addition, the knowledge about the use can be quite different from identifying and naming an object (e.g., Buxbaum et al. 2000). Finally, the patient was able to match a tactile experienced object to a picture, but produced an error rate of 16 % (84/100) when matching this tactile information to a spoken word.

D. Hillert, *The Nature of Language*, DOI 10.1007/978-1-4939-0609-3_10,

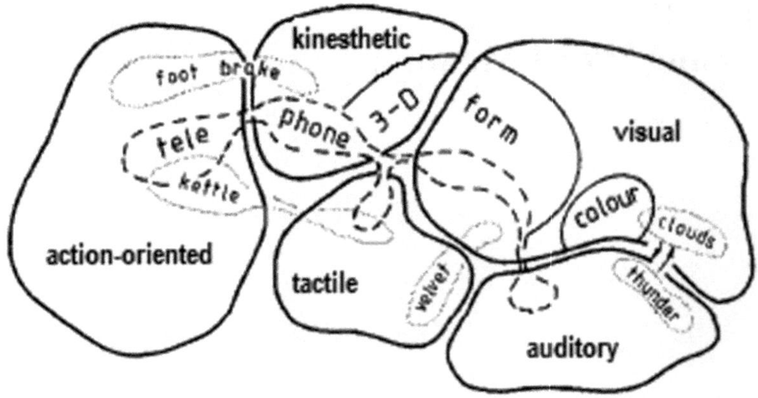

Fig. 10.1 The meaning of concrete words (object concepts) might be distributed across different sensory and motor domains. (Adapted and modified, © Allport 1985; © Elsevier Limited)

Such modality-specific disorders were taken as evidence for the view of a multiple conceptual system (Shallice 1987). Since, the participant had a comparable performance probability (0.68) for tactile object naming and matching tactile information to a word, it has been alternatively assumed that lexical access is impaired (Riddoch et al. 1988). Apart from this classical debate about the structure of the lexical conceptual system, the data can be also accounted for by an impairment of a unifying processing, which binds different experiences in a common conceptual system. In this case, the selective disorder would be attributed neither to a specific perceptual dimension nor to a central amodal conceptual impairment, but to unifying binding processes between different input data (Fig. 10.1).[1]

Let us briefly look at a few examples of category-specific disorders. The study by Hart et al. (1985) reported in the case of the aphasic patient MD, who used to work as a system analyst[2], a specific naming impairment for fruits (39 %) and vegetables (33 %). Also, MD was not able to produce the names for fruits and vegetables when their definitions were provided (2/10). In contrast, for items of other lexical categories, he had an error rate of only 3 %. For example, he was unable to name the picture of an orange or peach, but was able to name the picture of an abacus or sphinx. Selective naming disorders were also reported for other lexical categories including proper nouns, animals, country names, body parts colors, artificial objects, tools, letters, numbers, and concrete objects.

However, most behavioral studies favoring a multiple semantic systems account do not speak against an amodal semantic system. Most data can be also predicted by cascade processes within a single lexical conceptual system; that is, later processes

[1] Unification processes (also called binding processes) can be *per se* fractionated (Humphreys et al. 2009).

[2] Usually, the profession of the patient examined will not be mentioned in context of the neuropsychological evaluation. However, to some extent this information might be not completely irrelevant as the profession might indicate the person's lexical expertise.

can be activated before earlier processes are completed. In addition, these single-case studies often did not sufficiently control the stimulus material to consider more general lexical categories or particular perceptual computations (see for a critical review Hillert 1990, pp. 173–187; Hillert et al. 1994). In fact, if we accept compositional strategies at the computational level (not necessarily at the representational level) different dimensions interact to match the target concept associated with a particular name (lexeme). Some of these dimensions might be specifically degraded and may thus exacerbate the conceptual activation required for lexeme selection.

In fact, one study evaluated four structural dimensions (objects parts, internal details, object contours, variability of the representation) and their contribution to possible processing differences (Marques et al. 2013). They found that living things are structurally more complex than nonliving things. Living things have more object part and contours than nonliving things. Also, living things are less diverse across items, are visually less familiar, but their contours overlap greater than in nonliving things. Again, animals have more parts, fruits and vegetables more object contours, and nonliving things show less variability of the representation. Differences of the structural dimensions were also found at more specific lexical categories such as birds, fruits, insects, furniture, and vehicles. Thus, specific low score in one of these lexical categories compared to another one can be attributed to structural differences. As such Marques and colleagues interpreted their findings in line with a presemantic account of category effects. According to this presemantic account lexical category effects are affected by different task parameters including the required perceptual differentiation, duration, and position of stimulus or/and form of presentation.

Furthermore, object naming is a highly specific production task. Thus, based on this specific task condition we cannot conclude that comparable erroneous computations would occur during spontaneous speech or during an online processing task using the same stimulus material. Thus, it is essentially a methodological weakness, which let us doubt that these selective response patterns found in patients with different neurological disorders such as aphasia, dementia, or encephalitis support a modality-specific or sublexical conceptual system and argue at the same time against a unitary conceptual system. As a matter of fact, sensory- and language-specific access to conceptual representations represents a plausible working model.

In this vein, the model proposed by McClelland and Rogers (2003) seems to be a plausible approach. The idea of widely distributed lexical conceptual information is compatible with the findings that the output units of specific sensory information are located in distinct neural areas (Fig. 10.2). For example, when thinking about an action of an object the brain areas dedicated to this kind of information will be activated. In addition, representational units are proposed that unify object properties across different information types. These unifying representational units may be located in the temporal pole (TP). They are severely affected in the case of semantic dementia. The TP area may also be a repository of addresses or tags for lexical conceptual representations widely distributed. Finally, the model states that the medial temporal lobe (MTL) represents a complementary, fast learning system.

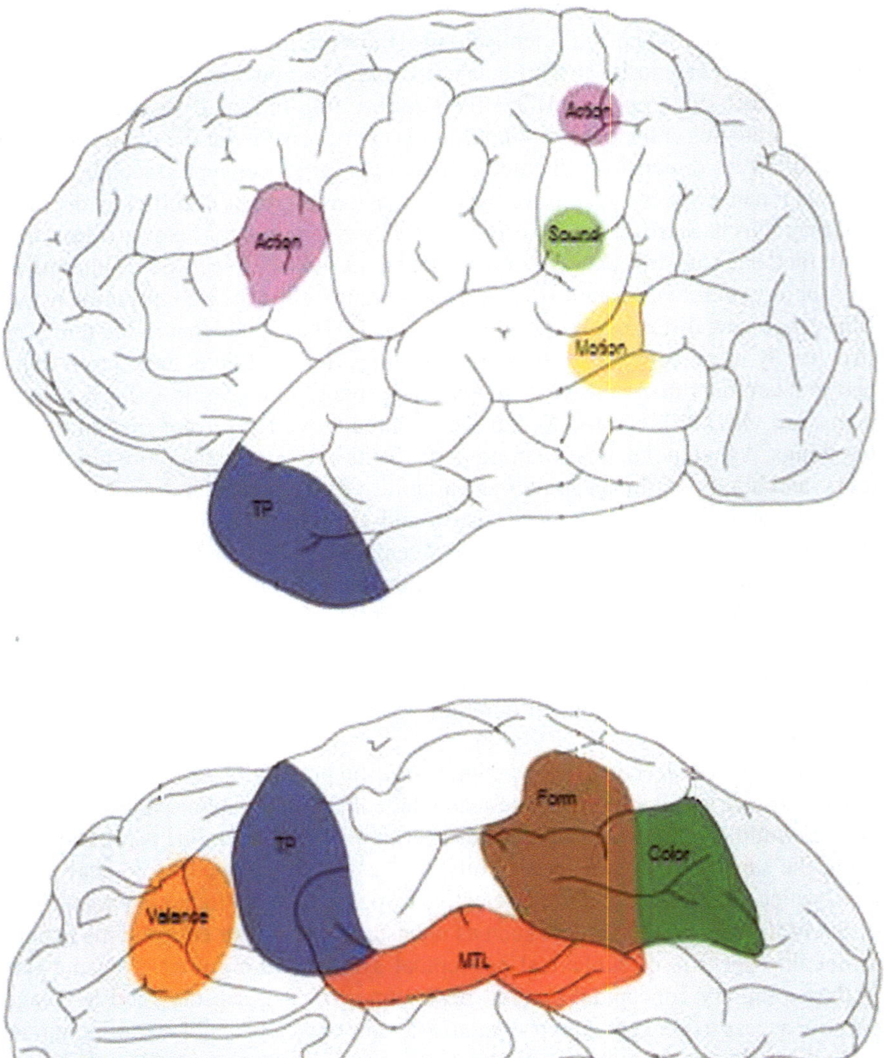

Fig. 10.2 Distinct sensory output units, unifying representational units in the temporal lobe (TP) and a fast learning system in the medial temporal lobe (MTL). (Adapted McClelland and Rogers 2003; © Nature Publishing Group)

A different approach examines the difference between object (nouns) and action concepts (verbs).

A meta-analysis of fMRI/PET activities on nouns and verbs by Crepaldi et al. (2013) does not support the assumption that verbs are mainly processed in the left frontal areas and nouns in temporal regions. In contrast, the findings show that nouns and verbs are processed in close spatial proximity in the predominately left-sided

Table 10.1 Verb and noun inflections for the three conditions A–C.

	Lead-in	Cue	Inner speech	Agreement	Morpheme	Potentials
A	(Instruction repeat)	"to walk" | "to think"	[wɑ́k] | [θíŋk]	N/A	V	~200 ms
		"a rock"	[rɑ́k]	N/A	N	
B	"Every day they"	"to walk" | "to think"	[wɑ́k] | [θíŋk]	V, 3. Pl., Pres.	V	~320 ms
	"That is the"	"a rock"	[rɑ́k]	N, Sg	N	
C	"Yesterday they"	"to walk" | "to think"	[wɒ́kt] | [θɒ́t]	V, 3. Pl., Pres.	V+ed | V'	~450 ms
	"Those are the"	"a rock"	[rɑ́ks]	N, Pl.	N+s	

V verb, *N* noun, *V_* null inflection, *V'* irregular verb, *Pl.* plural, *Sg* singular, *Pres./past* present/past tense

frontal–parietal–temporal regions. In addition, also neither the temporal course of verb and noun processing nor the temporal course of sublexical processing provide evidence for the assumption that these different linguistic units are computed by anatomically segregated structures. (Sahin et al. 2009). Although the authors argue for a subdivision of Broca's area with respect to different sublexical processes, the data speak for cascade processes involving different computational costs. Let us look at this study more closely. In this study, an intracranial procedure (LFP: local field potentials) was used to examine in three epileptic patients the temporal course of lexical production in Broca's area. Their study employed a phrasal completion task, in which an uninflected verb or noun (e.g., *to walk* or *a rock*) was displayed as a visual cue for the target word to be produced silently. The processing time for verbs or nouns increased in the following order (see Table 10.1): repeating (~200 ms) > covered (null) inflection (~320 ms) > overt inflection (~450 ms). The authors conclude: *"Broca's area is ... differentiated into adjacent but distinct circuits that process phonological, grammatical, and lexical information."* However, in the light of the meshing and incremental processes tested and of previous findings associated with the functional role of Broca's area, a more cautious interpretation is required.

The reported LFP results were collapsed across verbs and nouns. The findings indicate that the LFP method has limitations for the fine-grained examination of linguistic structures. The data are however to some extent compatible with differences found in priming, electrophysiological (scalp-surfaced event-related potentials), and fMRI, that is, lexical production involves different computational stages (e.g., Levelt et al. 1999; Friederici 2002). However, when considering the cognitive processes underlying the three conditions A–C, it is apparent that the temporal differences do not necessarily reflect sequential computations but different processes corresponding to different computational complexity. As mentioned before, the study examined three sublexical processes: (A) repeating, (B) null inflection lead in by a phrase and (C) overt inflection lead in by a phrase. Simplified, all three computations can be illustrated in (43) as follows:

(43) (A) Lexical recognition : phonological rehearsal

(B) Lexical recognition : phrasal integration : phonological rehearsal

(C) Lexical recognition : phrasal integration : morphemic adaptation : phonological rehearsal

The 200-epoch (A) requires copying the reading operation for silent rehearsal and matches well findings of behavioral tasks: one-syllabic words are recognized in a period of 75–100 ms, while the process of rehearsal takes place between 150–200 ms. The 320-epoch (B) adds another processing step as a lead-in phrase requires a congruent match between the target word and the phrase. A phrasal integration typically includes grammatical agreement and the generation of consistent syntactic–semantic structures. The phrasal integration process adds a response time of 120 ms. The 450-epoch (C) required in addition to the previous condition (overt/covert) inflection of the target word. This additional process, which involves the generation of a new word form (morphological adaptation), caused an additional response time of 130 ms. Thus, these cascade processes operate at the sublexical level and increase in complexity (A<B<C). Moreover, phonological, syntactic, semantic, pragmatic, and nonlinguistic computations engage Broca's area (e.g., Grodzinsky and Santi 2008; Hillert and Buračas 2009). These studies indicate that the cognitive architecture of Broca's area is multifunctional (*n*-divisions are possible). The question to be asked is: What is the common principle of those functions that correspond to the engagement of Broca's area? In fact, the storage hypothesis seems to be compatible with a wide range of data including the present LFP study (Stowe et al. 1998). An increase in cognitive complexity shifts working memory demands from posterior regions to Broca's area and speech is a domain of Broca's area to program articulation.

Let us return to the question how the human brain computes and represents lexical conceptual information. Functional neuroimaging data point also to cascade processes in object naming. For instance, if an object can be conceptually identified, probably by top–down processes, but not named by a patient, it can be the result of deficient perceptual information processes required to name an object. Although the concept has been accessed, recurrent perceptual information cannot be used for activating the corresponding lexeme (Humphreys et al. 1988; 1997; see also Perani et al. 1995; Martin et al. 1996). In this vein, we favor a general computational theory of lexical meanings (and sublexical information), which makes use of cascade processes in a distributed fashion. At the lexical conceptual level some research has argued that neuropsychological evidence would support an embodied account of lexical processing. However, sufficient evidence shows it is an account, which does not hold for all lexical phenomena. We emphasize that not all lexical meanings can be directly grounded by sensory-motor information. Spreading activations within an interactive neural net are reflected in bidirectional automatic activations between elements of the motor system and of perceptual, conceptual, and/or lexical elements (see for an excellent review Mahon and Caramazza 2008). Certainly, we associate sensory-motor experience with concrete as well as with abstract (symbolic) lexical concepts, but these grounded elements are only a part of a concept's representation. For instance, when asked to name a color, an action, or an object with characteristic visual features, it is possible that mediation by means of conceptual representations is not required (e.g., Caramazza et al. 1990). However, this does not exclude the fact that in other cases conceptual mediation is used (e.g., when the sensory experience of the experienced color is less typical and difficult to classify). Moreover, action words or action sentences activate relevant structures in the sensorimotor cortex. For example, the leg or the arm area will be activated, if the presented verb is respectively associated with the action of a leg or arm (Oliveri et al. 2004a; Buccino et al. 2005; Pulvermüller et al. 2005). However, before specific motor areas are activated, conceptual representations might mediate this process. Current methods may not provide the temporal resolution to measure this mediation process. It is the conceptual representation of a word or sentence that allows constructing meanings independent of our body experience. In general, there is no disagreement among researchers that lexical concepts are distributedly represented. The question is whether there are particular cortical hubs, which unify these different types of perceptual and conceptual information. Taking into account neuropsychological evidence from semantic dementia (fronto-temporal lobar degeneration), it has been claimed that the anterior temporal lobe should be considered as a "hub" for lexical conceptual information (Patterson et al. 2007). Others, who referred to aphasic disorders, consider the (bilateral) middle temporal lobe (MTG) as the hub for lexical semantic integration (Dronkers et al. 2004; Hickok and Poeppel 2007; Binder et al. 2009; Turken and Dronkers 2011). Binder's group discusses in their meta-analysis of 120 functional neuroimaging studies possible lexical conceptual subsystems. The SMG and the MTG, for example, would be involved in processing action knowledge. SMG is a portion of the parietal lobe and borders anterior the somatosensory cortex. Thus, it is suggested that this region stores abstract somatosensory conceptual knowledge acquired during complex motor learning rather than sensory-specific information as SMG is activated also for words with merely action meanings. The second area, the MTG, borders anteriorly the visual motion areas in the middle temporal lobe (V5). Here, it is assumed that visual attributes of actions are stored (Martin et al. 2000). Again, more abstract lexical concepts would be primarily processed in the left IFG (BA 47) and left anterior superior temporal sulcus. However, in considering a broader range of neuroimaging data, it turns out that lexical conceptual system consists of a large neural net involving multiple cortical areas. The trend shows that indeed stimulus- and task-specific demands determine the cortical subregions primarily active by default. The research reported reflects averaged activations across a relatively large number of study participants. It seems to be important, particularly in the domain of lexical concepts, that single-case neuroimaging approaches are more considered to draw further conclusions about how our different experiences shape our neural circuits.

10.2 Figures of Speech[3]

If you read a newspaper headline that includes the phrase *burned the midnight oil* you immediately may understand that it is related to studying long hours. This is possible because people have stored fixed strings of words in their long-term memory. Most people use a fairly large portion of such nonliteral fixed expressions to convey figurative meanings during everyday conversation. An expression such as *to live in an ivory tower*, for instance, cannot be understood by a literal–compositional analysis, but requires a figurative reading. Different cognitive process models have been proposed to describe how figurative meanings are accessed and contextually integrated.

Literal-independent accounts assume that access to figurative meanings occurs completely independent of the literal reading (e.g., Bobrow and Bell 1973; Gibbs 1980; Hillert and Swinney 2000; McGlone et al. 1994; Ortony et al. 1978; Peterson and Burgess 1993; Peterson et al. 2001; Swinney and Cutler 1979). In contrast, literal-dependent accounts, as the configuration hypothesis, claim that the literal reading will be suppressed in favor of a figurative interpretation after encountering a specific lexical configuration (e.g., Cacciari and Glucksberg 1991; Cacciari and Tabossi 1988; Tabossi and Zardon 1993, 1995).

(44) It's raining ca↓ts and dogs.

configuration point

However, even idiomatic expressions reflect a heterogeneous category and item-specific factors may determine how figurative meanings are accessed and contextually integrated (e.g., Cronk et al. 1993; Van Lancker Sidtis 2004). Thus, different comprehension processes may be involved when people access the figurative meaning of literally noninterpretable (e.g., *it rains cats and dogs*) or literally interpretable idiomatic expressions (e.g., *to break the ice*). It seems to be difficult to favor one idiom model over another, since they focus on specific aspects and their evidence stems from different experimental paradigms (see for a review Hillert 2008). However, these accounts share the discussion about the role of literal processing during accessing the meaning and integrating contextually the figurative interpretation. It is therefore an important factor that empirical approaches contrast ambiguous (literally interpretable) and explicit (literally noninterpretable) idiomatic expressions to obtain specific information about the computational steps involved in comprehension of figurative meanings. Moreover, as the graded saliency hypothesis predicts (Giora 1997), the factors familiarity and/or degree of figurativeness may influence strongly the interplay of literal and figurative meanings during the computation of figurative expression.

However, here we try to learn more about the neurological foundations of figurative language processing in general and about idiom processing in particular. In Laurent et al.'s (2006) ERP study, for instance, the N400 amplitude of the last words

[3] Parts of this section are taken from Hillert and Buračas (2009).

of salient idioms was reduced compared to less salient idioms, and in Mashal et al.'s (2005) fMRI study comprehension of novel metaphoric word pairs engaged significantly more regions of the right temporal lobe compared to conventional metaphoric (idiomatic-like) or literal word pairs. Moreover, it is unclear, which cortical areas are specifically engaged during idiom comprehension. Inconsistencies between the results of different studies are often related to the above mentioned factors.

For instance, early lesion studies on figurative meanings examined primarily metaphors rather than idiomatic phrases (e.g., Brownell et al. 1984; Brownell et al. 1990; Foldi 1987; Joanette et al. 1990; Van Lancker and Kempler 1987; Winner and Gardner 1977). These studies seem to indicate that the right-hemisphere plays a particular role in figurative processing. However, the fact that patients with right-sided lesions typically suffer also from visuospatial deficits has not been sufficiently controlled in these studies. Furthermore, lesion studies with aphasic patients indicate that idiomatic phrases might be processed left-sided or bilateral (e.g., Papagno et al. 2006; Papagno and Caporali 2007). The results of studies that used repetitive transcranial magnetic stimulation (rTMS) or functional neuroimaging techniques are however ambivalent. In healthy subjects, only left temporal rTMS disrupted string-to-picture matching performance (Oliveri et al. 2004b) and a PET study reported bilateral cortical activity for metaphor comprehension (Bottini et al. 1994; see also Nichelli et al. 1995; Subramaniam et al. 2013).

More recent studies applied fMRI to examine the neural correlates of figurative processing in metaphors (e.g., Ahrens et al. 2007; Rapp et al. 2004; Stringaris et al. 2007) and idioms (Lauro et al. 2008; Lee and Dapretto 2006; Zempleni et al. 2007). For instance, at the word level Lee and Dapretto's (2006) study used an adjective triplet priming paradigm (adapted from the subtest developed by Gardner and Brownell's 1986). The first two adjectives consisted of polar adjectives and the third adjective was either similar or opposite to the literal or figurative meaning of the second adjective (e.g., *hot–cold–chilly*). Subjects performed a lexical decision task on the second and third adjective of each triad. This lexical priming paradigm revealed reliable activity in the left fronto-temporal network when comparing the conditions "nonliteral vs. literal." In general, the authors conclude that only more complex figurative language may correlate with an increase of right cerebral activity (see also Rapp et al. 2004). Accordingly, when conventional metaphoric expressions were compared against literal sentences, Ahrens et al. (2007) found in a reading task increased right inferior temporal activity for conventional metaphors; however, anomalous metaphors engaged the bilateral fronto-temporal net. Again, in Zempleni et al.'s (2007) study ambiguous and explicit idioms were presented visually phrase-by-phrase. A bilateral activity in the IFG as well as in the MTG was reported. In Lauro et al.'s (2008) fMRI experiment, in which a written idiom-picture matching task was used, a specific activity was found in the bilateral fronto-temporal network for both ambiguous and explicit idioms. In using dynamic causal modeling to estimate the coupling of brain areas (Lee et al. 2006), Lauro et al. found in addition a specific bilateral activation for idioms in the medial prefrontal area. Finally, Mashal et al. (2008) reported that the literal interpretation of ambiguous idioms was supported by the right fronto-temporal circuitries, while the idiomatic reading engaged left-hemispheric regions.

The ability to understand irony (or sarcasm) involves listener's competence to understand the speaker's beliefs and intentions as a social–communicative function. (Here, we use the term *irony* in a broad sense, although it may refer to sarcasm in some cases.) Verbal irony is when the speaker uses his/her language to express the opposite of his/her thoughts. However, in contrast to lying, the speaker gives the listener some clues about the rhetorical expression. These clues can vary from an unusual intonation/stress, word choice, blinking, smirking to other nonverbal actions.

Only modern humans seem to be able to acquire this communicative ability. Accordingly, there is a substantial need to learn significantly more about the cognitive and neural mechanisms underlying the use of a theory of mind, world knowledge, and mental anticipation during language comprehension. Sperber and Wilson's (2001) relevance theory considers the intentional aspect of human communication, that is, another person's belief, but also the attribution of a belief about another person's belief. Thus, an ironic discourse is an excellent candidate for investigating the cognitive and neural second-order mechanisms required for assessing the speaker's beliefs (e.g., Winner 1988; Frith and Frith 2003). Grice's (1975) classical account implies that the listener revises the initial interpretation of an utterance only after a denotative-literal reading has failed. In contrast, this account has been challenged by psycholinguistic studies revealing that under specific conditions a nonliteral connotative reading will be directly accessed without the involvement of literal (de) compositional processes (e.g., Gibbs 1980; Hillert and Swinney 2000; Hillert 2004; Peterson et al. 2001). As discussed earlier, sentences expressing literal meanings are typically processed in specific anatomical subregions of the fronto-temporal net (e.g., Binder et al. 1997; Poldrack et al. 1999; Roskies et al. 2001; Bookheimer 2002; Gitelman et al. 2005). In contrast, figurative aspects of language processed in a more distributed manner in the left as well as in the right temporo-parietal region and in the left middle and superior frontal gyrus (MFG and SFG) (e.g., Kempler et al. 1988; Tompkins et al. 1992; Kircher et al. 2001; Papagno et al. 2002; Paul et al. 2003; Just et al. 2004; Lee and Dapretto 2006; Oliveri et al. 2004b; Rapp et al. 2004; Sotillo et al. 2005).

The findings indicate that linguistic pragmatics requires the interaction between right and left temporo-parietal regions as well as activities in the left frontal gyrus, while core language processes rely mainly on specific cortical areas in the left inferior frontal gyrus (IFG; particularly BAs 44/45 and 49) and in the left temporo-parietal region (particularly BA 22, the anteroventral temporal lobe (avTL) and angular gyrus, AG). Only a few studies so far examined the neural substrate of irony comprehension and its clinical relevance. More recent studies reported consistently mPFC (about BA 10) activation related to mentalizing (e.g., Frith and Frith 2003; Amodio and Frith 2006), reflecting on own emotions (Gusnard et al. 2001) and on character traits (Macrae et al. 2004). Also, Gilbert et al. (2007) reported a subdivision within the mPFC, that is, the medial rostral PFC (mrPFC) seem to be involved in attention selection between perceptual and self-generated information while a caudal-superior region of the mPFC appears to be more involved in mentalizing, that is, in reflecting on the cognitive states of another agent (cf. Gilbert et al. 2006).

Only a few fMRI studies examined the neural basis of irony comprehension (Eviatar and Just 2006; Uchiyama et al. 2006; Wakusawa et al. 2007; Shibata et al. 2010). Eviatar and Just (2006) used three-sentence stories to compare comprehension of literal, metaphoric, and ironic statements. Depending on the type of meaning examined, they found different inter-hemispheric sensitivity: As compared to metaphoric statements, ironic statements produced significantly higher activation levels in the right S/MTL. Uchiyama et al.'s (2006) fMRI study used a scenario-readig task for recognizing sarcastic statements and found activation in the left temporal pole, the superior temporal lobe and in the mPFC. Again, Wakusawa et al. (2007) matched utterances to photos of common situations to examine understanding of implicit meanings with respect to a particular social context. They reported specific right TP activation for ironic processing, but also an involvement of the medial orbitofrontal cortex, an area that is also involved in decision-making (Kringelbach 2005). These results are comparable to tasks requiring "mentalizing" (e.g., Brunet et al. 2000; Gallagher et al. 2000). Finally, in Shibata et al. (2010) study echoic utterances were used as ironic statements. The condition "ironic minus literal statement" revealed increased cortical activation in the right medial prefrontal cortex, the right precentral sulcus and the left superior temporal sulcus. In general, the present brief review of the neuroimaging data indicates that pragmatic language engages a broader cortical network than literal language. Meta-linguistic pragmatic computations (mentalizing) that are required for irony comprehension seem to engage in particular the right-sided cortical regions including mPFC, the superior temporal lobe, and the TP in addition to the classical left-sided language net.

It remains an open question to what extent stimulus types and form of presentation contributes to the activity in terms of magnitude and spatial–cortical extent. Hillert and Buračas (2009), therefore, conducted an fMRI study to examine the neural correlates of spoken comprehension of idiomatic expressions in a naturalistic fashion simulating conversational linguistic behavior (see also Tompkins et al. 1992; Hillert 2004, 2008, 2011). To examine the extent to which a literal analysis is involved in idiom comprehension, we presented sentences that included ambiguous (literal and idiomatic meanings) and unambiguous, explicit idiomatic sentences. Our first hypothesis stated in light of previous findings that both types of sentences expressing a literal or idiomatic meaning will activate the left fronto-temporal net. Our second hypothesis was that ambiguous sentences allowing a literal or a figurative reading will involve different left-sided cortical regions as compared to explicit idiomatic sentences. Previous fMRI study did not report different cortical activations for comprehension of explicit and ambiguous idioms (Lauro et al. 2008; Zempleni et al. 2007a). However, both studies used a relatively slow-paced functional paradigm, which involved phrase-by-phrase judgments or phrase-to-picture matching tasks. Here we used a rapid auditory sentence decision (rASD) task to simulate comprehension of familiar and short idiomatic expressions during online spoken sentence processing. We believe that rASD may be more sensitive to different levels of linguistic processes than tasks which do not impose temporal constraints.

Comprehension of sentences with standing ambiguities, that is, context of any kind does not provide sufficient clues to resolve the ambiguity, may involve parallel computations of the literal and idiomatic meaning. The parser may anticipate a literal Agent–Theme structure at an early stage by providing an argument grid with Theme as a placeholder. However, the co-occurrence of a specific verb–noun combination (V–N: *break-ice*) generates an additional idiomatic parse. Since context information does not bias a particular reading, only subjective preferences, which might reflect common sense, can be used to solve a standing ambiguity. In the case of explicit idiomatic expressions different scenarios are possible depending in particular on the length and familiarity of the expression (e.g., Gibbs 1980; Hillert and Swinney 2000; Tabossi and Zardon 1995; Titone and Conine 1994). Since the literal reading is impossible or highly implausible, the figurative meaning may be intrinsically selected even before the complete expression has been perceived. This leads to suppression or deactivation of the implausible literal interpretation. Based on priming data mentioned above, we assumed that early parallel computations take place even when explicit idiomatic expressions are parsed. Since early parsing does not resolve standing ambiguities, we predicted that comprehension of ambiguous idiomatic expressions may activate a broader left-sided cortical net than comprehension of explicit idiomatic expressions.

We prepared as stimulus material short sentences with a maximal length of 1.8 s, which all consisted of a grammatical subject followed by a verb phrase (VP). While the Subject was kept semantically abstract and neutral, the VP differed in terms of ambiguity, figurativeness, and meaningfulness. Accordingly, the experimental stimuli included explicit idiomatic sentence (e.g., *He was shooting the breeze*) and ambiguous idiomatic sentences (e.g., *The woman held the torch*). As control stimuli we used literal sentences (e.g., *He met her in the new mall*) and implausible nonsense phrases violating basic common sense (e.g., *He ate the green wall*). Participants were asked to rate 250 short sentences on a seven-point scale presented in a random order according to their degree of figurativeness (Fig. 10.3). A rating of 1 was regarded as nonfigurative, that is as literal, and 7 as highly figurative. We asked our participants to rate the implausible phrases as zero.

After reclassification of the sentences with respect to the lexical categories of interest, we found a trend for the expected patterns. However, the subjective ratings were less clear-cut as assumed per definition. The most homogenous responses were found for literal phrases followed by implausible phrase, of which some phrases were considered to have a weak figurative meaning. The intersubjective variability was higher for both idiomatic phrases, whereas the maximal score of figurative readings was only reached by explicit idiomatic phrases. All comparisons between these phrasal types where highly significant. In particular, the contrast between ambiguous and explicit idiomatic phrases let us assume that raters take into account the possibility of a literal reading when rating the figurativeness of ambiguous idioms (Fig. 10.4).[4,5]

In testing ten young healthy volunteers, the group analysis of the event-related fMRI scans addressed cortical regions that covaried with the above mentioned

[4] The stimuli were controlled for other variables such as familiarity (see for details Hillert and Buračas 2009).

[5] This is also the brain that wrote this book.

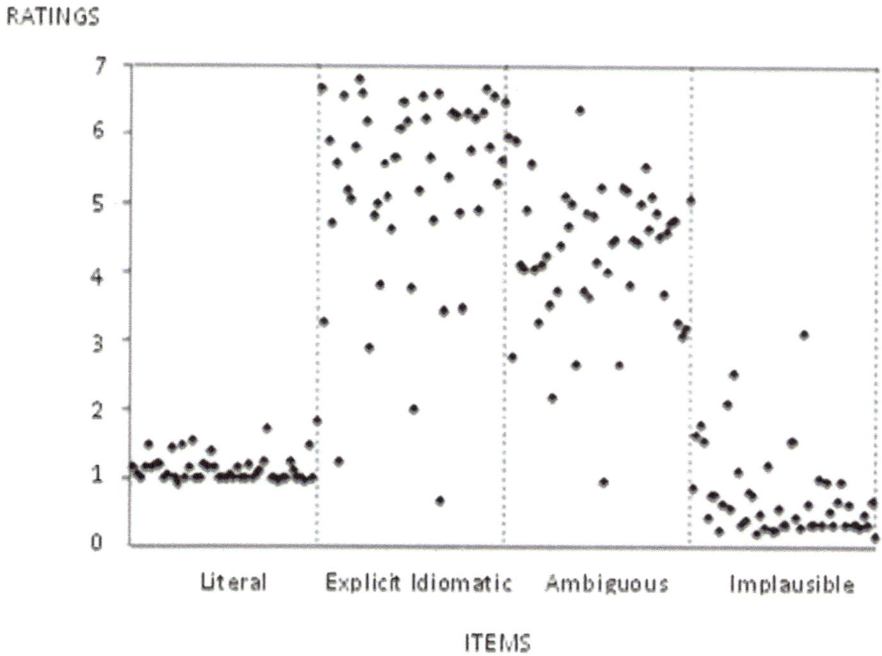

Fig. 10.3 Seven-point ratings of figurativeness with respect to different lexical categories. The rating of "zero" was reserved for implausible meanings. (Adapted and modified, Hillert and Buračas 2009; © Taylor & Francis)

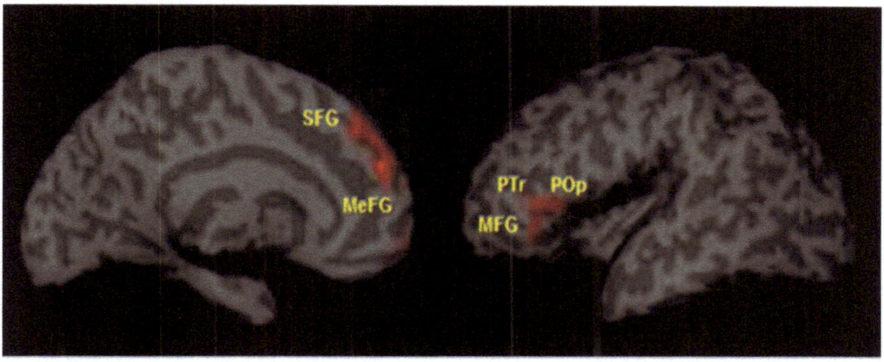

Fig. 10.4 FMRI analysis of figurativeness ratings projected on a single mid-/sagittal view of an inflated left-hemisphere reveals two different activity clusters (see text for details).

sentence figurativeness. The cluster analysis revealed two different areas of left-sided activations: One area involved the left pars triangularis (PTr) and pars opercularis (POp), that is Broca's area (BAs: 44 and 45) and the middle frontal gyrus (MFG; BAs 11 and 47); the other area included the superior frontal gyrus (SFG; BA 8) and the medial frontal gyrus (MeFG; BAs 8 and 9). The computation of

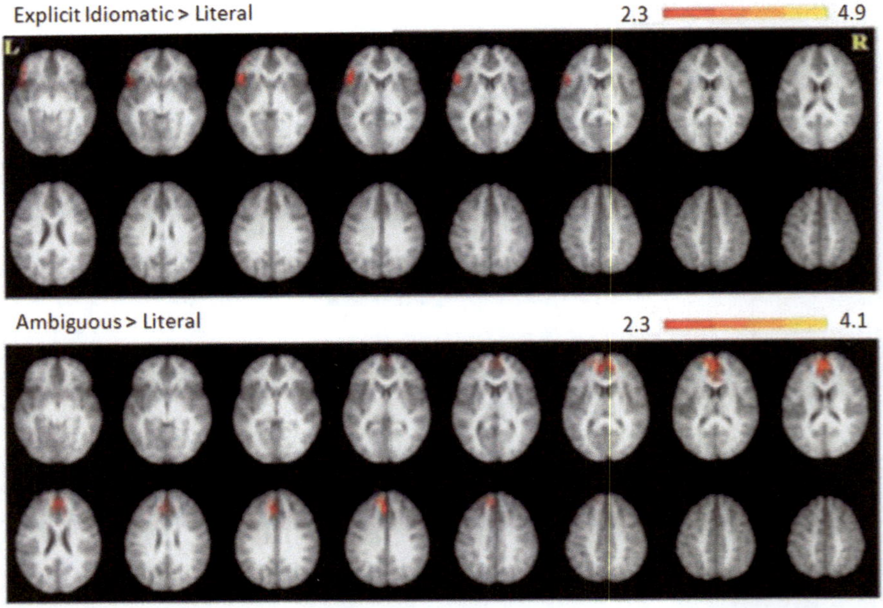

Fig. 10.5 Comparisons between idiomatic and literal phrases with z-scores, Z (Gaussianised T/F) statistic images were thresholded using clusters determined by Z score > 2.3 and a (corrected) cluster significance threshold of $p = 0.01$, Worsley et al. 1992. (Adapted and modified, Hillert & Buračas 2009; © Taylor & Francis)

idiomatic meanings seems to be mainly a domain of the left prefrontal cortex. No evidence was found for a right-hemispheric contribution.

In the second step, we were interested to learn whether these two clusters are related to particular idiomatic expressions. Thus, we analyzed post-hoc at $p < 0.01$ both activation clusters by including all pairwise comparisons of the four sentence types. The comparison "explicit idiomatic > literal" evoked clusters in the left PTr and POp as well as adjacent areas (BA 46 and 47), and the comparison "ambiguous idiomatic > literal" revealed cortical activity of the MeFG and the SFG in the vicinity of the frontal midline with left-sided prevalence, but no significant activity in Broca's area (see Fig. 10.5).

Our first analysis of figurativeness revealed two different left-sided cluster activities: one cluster was evoked in Broca's area and the MFG, the other cluster in the SFG and MeFG. The cluster activity found for figurativeness seems to overlap to some extent with the cluster activity found here for the contrasts between "explicit idiomatic > literal" and "ambiguous idiomatic > literal." Thus, the first cluster activity that covaried with figurativeness in Broca's area and vicinity (BAs 44, 45, 47, 11) may refer to the cluster activity found for the contrast explicit idiomatic > literal (BAs 44, 45, 47). Both clusters involve PTr and POp and adjacent areas. The second cluster activity that covaried with figurativeness (BA 8, 9, 32) seems to correspond to the cluster activity found for the contrast "ambiguous > literal" (BAs 8–10, 32). These clusters, primarily left-sided, include the SFG, the MeFG, and the cingulate

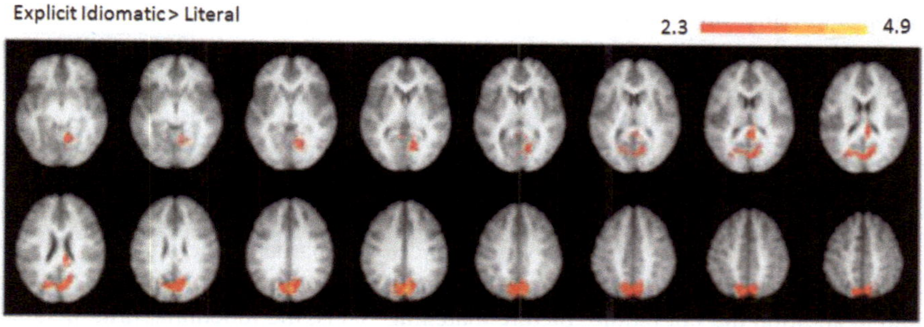

Fig. 10.6 Comparisons between literal and explicit idiomatic phrases reveals the involvement of the occipital and parietal lobe. (Adapted and modified, Hillert & Buračas, 2009; © Taylor & Francis)

region. Thus, the results of both fMRI analyses revealed approximately identical results supporting the account that in particular the left prefrontal cortex is involved in processing figurative meanings of idioms.

We reported furthermore another interesting result. The contrast "literal > explicit idiomatic" revealed left-sided peak activity in the occipital lobe (cuneus: BA 17) and of the medial parietal lobe (Fig. 10.6).

The literal phrases were used as control phrases, but why did they evoke activities in the visual cortex as compared to explicit idioms? Again, why was this effect not generated by the comparison with ambiguous idioms? A plausible interpretation of this result is that a compositional strategy may activate in a more elaborative fashion image to facilitate the comprehension process. The contrast between idiomatic and literal phrases does not show this effect as fixed expressions do not evoke deep semantic processing but are computed on the fly. In this vein, the comparison between "literal > ambiguous idioms" may not be significantly different as both involve a compositional strategy in contrast to the comparison "literal > explicit idioms." This comparison probes computations at both end points of a continuum from unambiguous fixed expression ready to be computed in a highly automated fashion to literal phrases requiring a compositional strategy. The differences between these computations rather than the occurrence of figurative meanings *per se* may have produced this significant effect by evoking the involvement of the cuneus, an area typically not involved when computing fixed expressions. Moreover, it is interesting to note that parietal lobe activities were found in addition to those found in the cuneus. The parietal lobe is the interface for integrating different kind of sensory information and might represent one important site for consciousness. A compositional computation requires to some extent more awareness than automated processes. As such the results match the account that it is not the type of meaning or content that drives the involvement of a particular cortical area (at least, if content is not sensory-driven), but the type of computation including costs associated with particular linguistic information. The evolution of the parietal lobe seems to have provided *Homo sapiens* with the ability to compute meanings across different domains, a hypothesis we discuss in a different subsection below.

Overall, the findings reported confirm only in part from a neurological viewpoint our first hypothesis as such that the relevant prefrontal regions are subregions of the left fronto-temporal network. The results seem to be most compatible with Oliveri and colleagues' (2004b) rTMS findings, and Lee and Dapretto's (2006) fMRI data. Oliveri et al. reported a particular role of the left temporal lobe in idiom processing. However, in this study a sentence–picture matching task was used and image processing engaged more posterior regions in proximity of the occipital lobe as compared to rASD task condition. For the figurative interpretation of lexically biased adjectives, Lee and Dapretto found left-sided fronto-temporal lobe activities. The task, *per se,* involved "shallow demands" compared to the study of Oliveri and colleagues since the stimuli were presented isolated, without sentence context, and in addition did not require picture matching. Thus, the functional task included comprehension of literally and figuratively interpretable adjectives at the lexical level. The reason why we found in our study cortical activities primarily in the left prefrontal cortex may be therefore related to the generation of syntactic structures involved in idiom comprehension.

Our current findings, however, clearly differ from those of other functional neuroimaging studies on idiom comprehension that support a bilateral model (Lauro et al. 2008; Mashal et al. 2008; Zempleni et al. 2007a). Again, we believe that these diverging results partly depend on different cognitive demands associated with the specific experimental paradigms used. For instance, Zemplini and colleagues used lead-in sentences to bias a literal or a figurative interpretation of the idiomatic expressions and presented the stimuli phrase-by-phrase. Again, in Mashal et al.'s study, participants were instructed before the presentation of a stimulus set to interpret the sentences either literally or figuratively, and Lauro and co-researchers examined idiom comprehension with a four-choice sentence–picture matching task. Both studies were contextually more demanding than the current rASD study since we did not present biasing sentences as lead-in or pictorial information. Additionally, our study used an event-related design, but all three studies mentioned above presented the stimuli in blocks. A block design may influence participants' comprehension strategies as such that repetitive computations may facilitate semantic analysis. We speculate that an increase of temporal lobe activity might be related to a decrease in prefrontal lobe activity as it can be observed in repetition priming (e.g., Demb et al. 1995; Wagner et al. 1997). In general, we did not find support for our second hypothesis that ambiguous idioms significantly engage a larger cortical net than explicit idioms. In terms of cluster size, the spatial extent for ambiguous and explicit idioms was comparable when respectively contrasted with literal sentences. However, the comparison between ambiguous idiomatic and literal phrases revealed cluster activity in the left but to some extent also in the right SFG and MeFG, that is, two out of six local cluster maxima were right-sided. Although, the cluster activity for this contrast is located in the vicinity of the frontal midline, more left-sided activity can be reported.

The picture changes when we briefly extend our discussion to metaphor processing. Bottini et al.'s (1994) PET study seems to have favored a right-hemispheric account of novel metaphor comprehension. Mashal and colleagues' (2005) study used Hebrew written word pairs and reported that in contrast to literal pairs novel metaphoric word pairs were mainly processed by the right homologue of Wernicke's area and conventional metaphoric (idiomatic-like) word pairs in Broca's area. Again, the findings of Ahrens et al.'s (2007) fMRI experiment with Mandarin Chinese sentences, however, are more ambivalent. Compared with literal sentences, they reported for conventional metaphors, a slight advantage for the right inferior temporal gyrus, and for anomalous (novel) metaphoric sentences, bilateral activities of the fronto-temporal network. In contrast, two silent reading studies during MRI scanning revealed a different cortical activation pattern for metaphors in comparison to literal sentences: In Rapp et al.'s (2004) study, reading of German sentences with novel metaphors engaged activity in the left IFG and in the middle and inferior temporal lobe, and in Stringaris et al.'s (2007) study nonsensical and (relatively conventional) metaphoric sentences recruited in particular BA 47 of the left IFG. The above-mentioned functional neuroimaging studies on metaphor processing differ largely with respect to the stimulus material used in terms of "metaphoricity," linguistic typology, and task condition. Such inhomogeneity or variance of stimuli and computational demands may be the reason for these inconsistent findings. Compared with idioms, metaphors seem to engage a broader cortical network. One reason may be that comprehension of the active metaphors we live by typically generates mental images and requires semantic frame-shifting. Comprehension of idiomatic expression, however, relies typically on syntactic shifting or suppression to parse the alterative figurative meaning. Supported by our figurativeness analysis, our fMRI data revealed that explicit idiomatic phrases engage Broca's area and adjacent regions (BAs 46 and 47), and ambiguous idiomatic phrases recruit primarily the left SFG and MeFG when compared to literal sentences. This result indicates that ambiguous idiomatic phrases are processed differently from explicit idiomatic phrases.

Functional neuroimaging studies, which examined lexical ambiguity resolution, support a bilateral contribution (e.g., Mason and Just 2007; Zempleni et al. 2007b), and lesion studies indicate that the prefrontal cortex plays a particular role in ambiguity resolution (e.g., Copland et al. 2002; Frattali et al. 2007). We examined in our study standing ambiguities, that is, no discourse or pragmatic context was provided to trigger disambiguation. On the one hand, this might be the reason why we did not find a direct involvement of right-sided cortical areas; on the other hand, it might explain why explicit idiomatic phrases were processed differently from ambiguous idiomatic phrases. Explicit idiomatic phrases specifically evoked cortical activities in Broca's area when contrasted with literal sentences. A critical function of Broca's area (BAs 44 and 45) seems to be related to rehearsing linguistic information for inner and/or overt speech and parsing complex phrases such as noncanonical sentence structures (e.g., Caplan 2001; Caplan et al. 1998; Grodzinsky 2000; Grodzinsky and Santi 2008; Kaan and Swaab 2002; Mason et al. 2003). However, all our plausible sentences that we examined possessed a canonical SVO (subject–verb–object) structure. We assume, therefore, that Broca's area is involved in any kind of cognitive computation for selecting the appropriate parse. This parsing mechanism involves also the suppression of an irrelevant syntactic analysis (e.g., the literal default parse in case of explicit idiomatic expressions). Moreover, we assume that standing ambiguity (when contrasted against literal sentences) did not engage

Broca's area as no contextual cue was provided to select a particular interpretation. The present fMRI results do not provide information as to whether a specific cytoarchitecture associated with the subregions of Broca's area is primarily engaged during idiom comprehension (see Dapretto and Bookheimer 1999). However, the data confirm that explicit idiomatic and literal sentence parsing is a domain of Broca's area involving syntactic–semantic rehearsal operations.

In general, the prefrontal cortex comprises a collection of interconnections with different brain areas to synthesize external and internal information. It is highly multimodal in nature and is suggested to reflect the neural substrates for processing abstract information (Miller 1999; Miller and Cohen 2001; Miller et al. 2002). To search for pragmatic or contextual cues, this prefrontal region may maintain alternative interpretations to direct information exchange with knowledge traces stored throughout the cortex. The data presented here show that idiomatic expressions are automatically processed in different cortical areas, which overlap only partially with the classical fronto-temporal language net. As for idiom comprehension, we found evidence for search processes primarily in the left SFG and MeFG and for parsing operations in Broca's area. These cognitive computations differ from those left medial parietal activities involved in comprehension of canonical and unambiguous literal sentences. It might be possible, therefore, that parietal–temporal activities related to the literal reading of an idiomatic expression are suppressed by those prefrontal modulations, which generate a figurative reading of that expression. The data presented by Hillert and Buračas strongly support a model of spoken idiom comprehension that engages specific areas of the left prefrontal cortex in association with certain idiom types. Our findings provide further evidence for a multifunctional dynamic cytoarchitecture that supports a parallel cognitive architecture of language processing.

To exemplify this idea, ambiguous idioms and explicit idioms both engage a literal and a figurative parse. However, both types of syntactic analyses do not directly correspond to a single specific cortical area since both idiom types activated different cortical areas. Linguistic computations seem not to be limited to specific cortical areas. Instead, the underlying (de)activation patterns between different computations determine which cortical areas are more active than others. Because of computational complexity both idiomatic structures examined appear to be mainly processed in the left prefrontal cortex rather than in posterior regions. The involvement of Broca's areas in the case of explicit idiomatic expressions may be related to rehearsal operations rather than to particular linguistic computations. Our research lets us assume that language processing involves intrinsic decisions which seem to operate throughout the cortex in a parallel fashion. The fMRI data indicate that unfamiliar or novel lexical meanings seem to engage a larger cortical net (including the right cortex) as compared to familiar or conventionalized lexical meanings. That is, the degree of composition required to process figurative meanings seems to be associated with a larger cortical net. This interpretation is in line with a meta-analysis of 22 fMRI studies (354 subjects) indicating that additional right-hemispheric activities are limited to novel metaphors (Bohrn et al. 2012; see also Rapp et al. 2012).

We present here an evolutionary account of figurative processing, which may have its origin in the development of the MNS—the mirror neuron system (Pineda 2009). This MN circuit responds to both, when we perform an action and when we observe an action. Human neuroimaging studies indicate that the observation of actions involved are variety of different cortical areas including the occipital, temporal, and parietal areas. However, two (bilateral) areas are primarily involved as they are mainly associated with motor functions: (a) the rostral part of the inferior parietal lobe (IPL) and (b) the posterior part of the IFG along with the lower part of the precentral gyrus (Buccino et al. 2001; Rizzolatti and Craighero 2004). The IPL, also known as Geschwind's territory, is located behind the lower part of the postcentral sulcus and can be divided into the supramarginal and the angular gyrus (see also Fig. 10.6). Again, the angular gyrus borders the superior and middle temporal gyrus. As compared to the cortex of the Neanderthals or even Homo erectus (including Homo ergaster), the entire parietal lobe is significantly enlarged in H. sapiens (Bruner 2010; Kaas 2013). However, the frontal lobe of Neanderthals and modern humans is wider as compared to H. erectus. In H. erectus, the brain's maximal width can be found at the temporal lobe and between the temporal and parietal areas in Neanderthals and at the parietal region in modern humans (Bruner 2004). Thus, we can assume that the parietal lobe region may have been less elaborated in *H. erectus* than in anatomically modern humans. It is highly speculative, as mentioned before, to take fossils as evidence for the ability and/or use of symbolic communication (Fig. 10.7).

However, in considering the estimated anatomically changes of the parietal region in Neanderthals, H. erectus and modern humans, we assume that the integration and use of multi-sensory information gradually increased and may have led to qualitatively new behavioral dimensions. The parietal lobe can be considered an interface between the prefrontal cortex and posterior parts including the temporal and occipital lobe integrating multi-sensory information and cross-domain knowledge. It is also assumed that the partial lobe is the primary seat of (self-) consciousness.

Everything said here remains speculative, but a model might be plausible that predicts the use of figurative meanings to modern humans. As we favor the view of a gradual genetic and social evolution of communicative abilities, some basic elements of the ability for nonliteral usage might have been already present in H. erectus. The play of words and the creation of novel linguistic forms is a social product, but the neurobiological conditions scaffolding the full range of figurative language as known to us today might be present only in anatomically modern humans. Lesions to the IPL, including the supramarginal gyrus and angular gyrus, results in affective language disorders related to metaphors, humor or irony comprehension, and in particular left IPL lesions cause ideomotor apraxias. The characteristics of an ideomotor apraxia, originally known as an ideo-kinetic disorder (Liepmann 1905; Goldenberg 2003), include the inability to understand or imitate gestures or execute common actions such as making tea. These actions typically consist of several steps requiring a certain sequence in time and space. It is to assume that along with the increase of the functional role of parietal lobe areas in the human lineage attentive, conscious computations and the ability of abstraction (including mathematics)

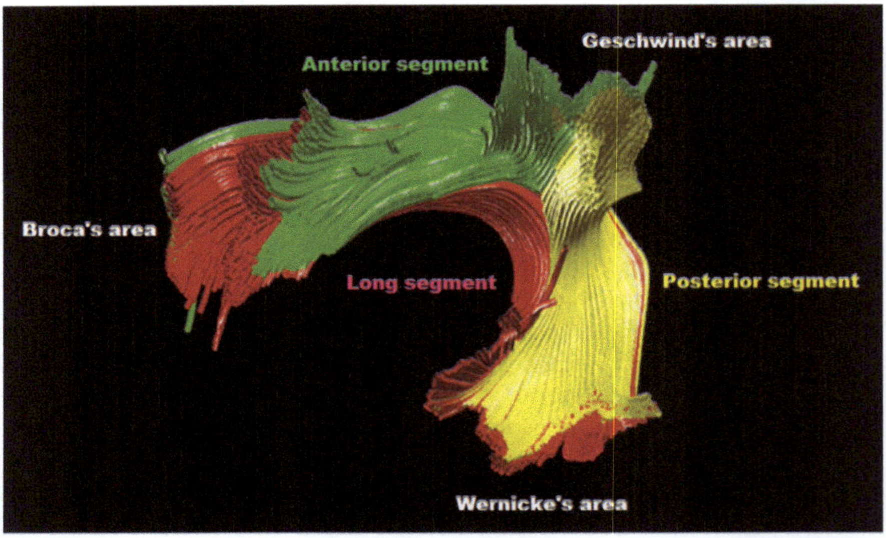

Fig. 10.7 Tractography reconstruction of the arcuate fascilicus (AF) connecting Broca's and Wernicke's area. The long segment represents the direct classical tract of the AF (marked in red). The anterior indirect segment marked in green runs between Broca's area and the inferior parietal cortex (Geschwind's area); the posterior indirect segment marked in yellow runs between Geschwind's area and Wernicke's area. (Adapted and modified, Catani et al. 2005; © 2005 American Neurological Association)

increased as well. The ability to use cross-domain and cross-sensory information for creating new meanings may be closely related to that what we commonly refer to as consciousness. In particular, artists such as painters, poets, or musicians may rely on this cortical function. In this vein, the takete/bouba effect reflects cross-sensory computations. In a Spanish study by Köhler (1929), a strong preference was found for associating respectively the syllabic combinations *takete* and *bouba*[6] to particular shapes (see Fig. 10.8). The reader may be certain, which sound form goes with which shape—yes, and this was/is indeed the case, also for toddlers (Maurer et al. 2006; see also Ramachandran and Hubbard 2001).

Synesthesia can occur between any two senses, and 60 different types of synesthesia have been reported. They represent a form of atypical neural wiring associating information between two sensory modalities, which are typically not associated. It seems to be genetically based condition, not necessarily linked to a specific synesthesia, as it runs in families. To name a few: numbers or letters are perceived always in specific colors (e.g., A tends to be red), numbers are seen as points in space (which lead to superior memory skills), sounds like voice, music, or context sounds are experiences triggering different colors, or ordinal numbers, days, months, and letters receive a personification.

[6] These are Spanish sound patterns as the participants were Spanish speakers.

Fig. 10.8 You can tell which shape goes with which sound form!

Calkins (1893) notes:

> "T's are generally crabbed, ungenerous creatures. U is a soulless sort of thing. 4 is honest, but … 3 I cannot trust … 9 is dark, a gentleman, tall and graceful, but politic under his suavity."

Neuroimaging studies reported that in particular the color center V4 (also called V8) is involved in color synesthesia (Nunn et al. 2002; Hubbard et al. 2005). V4 is located in the fusiform of the temporal lobe and is part of the ventral stream that connects to the inferior temporal cortex. However, the data projected from the primary visual cortex (BA 17) to the fusiform will be further processed to other[7] cortical areas for specific purposes. Thus, although fMRI studies reveal a significant role of V4 in the case of color synesthetes, further processing steps may involve the temporal–parietal regions. For instance, as numbers and colors are identified within the fusiform, the occurrence of a number–color synesthesia might be caused by an atypical cross-wiring within the fusiform (Ramachandran and Hubbard 2001). However, the cause of an atypical cross-wiring might also arise more downstream in the vicinity of the angular gyrus. Depending on the type of sensory input, the causes of an atypical cross-wiring may involve different projections, which all lead to the temporal–parietal–occipital junction—also known as TPO.

The increase of the parietal lobe (or TPO) area in the human lineage may have emerged from the demand for symbolism to enable communication not only based on actual sensory experiences but also which rely on imagination and simulations. Unfortunately, fossil evidence cannot provide clues about the evolving parietal cortex in the human linage. However, some regions of the posterior parietal cortex, in particular the angular gyrus, seem not to have homologues in macaque monkeys (Orban et al. 2006). Measuring the relative difference between functionally homologous areas V5 in the extrastriate visual cortex (motion sensitive area, also known as middle temporal— MT) and A1 (the primary auditory cortex) indicate a large expansion of the human cortex between these areas (Fig. 10.9). The outcome of several comparative techniques identified a zone in the posterior part of the human parietal lobe,

[7] Often the term "higher" cortical area is used in the literature. To avoid any bias toward the functional role of a cortical area, we prefer to use more neutral terms when contrasting the functional roles of different cortical areas.

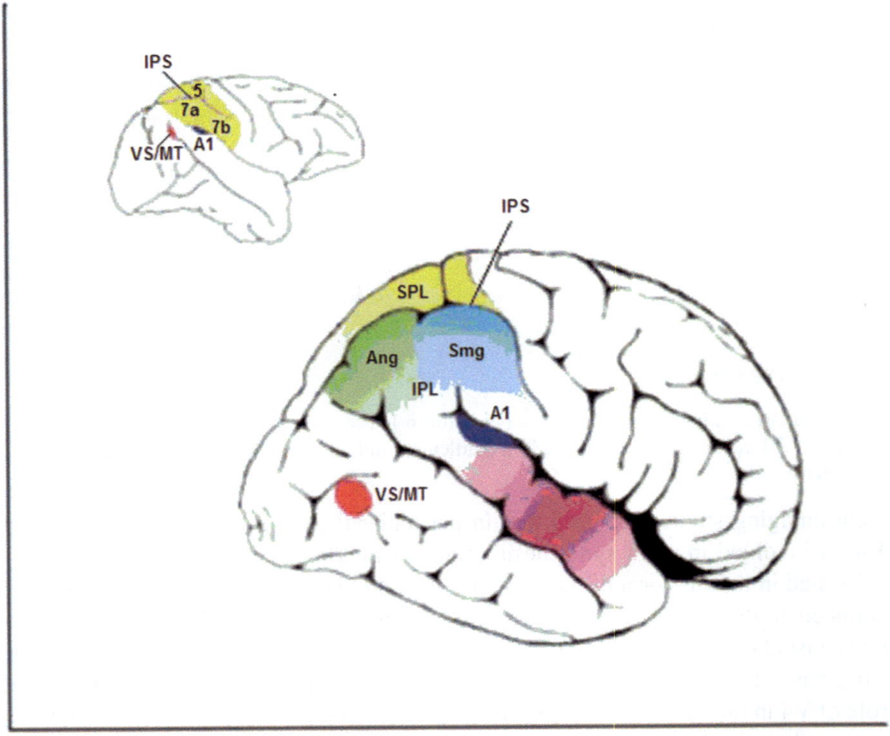

Fig. 10.9 Parietal cortices in modern humans (front) and macaques (behind). In modern humans, the intraparietal sulcus (IPS) divides the superior parietal lobe (SPL) from the inferior parietal lobe (IPL). Since the homologues in macaque (areas 5 and 7a/b) confine to the human SLP, the IPL in modern humans are thus considered as new cortical areas. (Adapted and modified, Husain and Nachev 2007; © Elsevier Limited)

which seems not to have a direct homologue in the macaque brain (Caspers et al. 2006; Husain and Nachev 2007).

We do not know how the brain of modern humans evolved, but the assumed gradual increase of the posterior parietal lobe may have significantly contributed to the origin of the human language faculty. Did ancestral humans develop linguistic syntax based on the assumption that rhythmic patterns experienced in single sensory modality such as tool use or vocalizations were represented modality-independent across specific modalities? It is a plausible hypothesis as linguistic syntax has structures that partially overlap with music syntax. In favor of this assumption speak also lesion studies. In contrast to monkeys, substantial hemisphere-specific functional differences can be observed in modern humans. While lesions in the right inferior parietal lobe and temporal–parietal junction typically cause neglect, lesions in homologous areas of the left hemisphere often cause apraxia. Some left IPL patients, however, show also right-sided neglect. Such hemisphere-specific parietal lobe differences have not been reported for monkeys. Further research is required to draw more precise conclusions about the role of the parietal lobe in the evolution of language and in language processing.

The development of the mirror neuron system might have played a particular role in the evolution of the human language faculty. To sum up: In macaques, mirror neurons have been directly identified in F5 (ventral premotor cortex) and PF/PFG (inferior parietal lobe). In modern humans, the IFG, IPL and STS will be activated when actions are observed and executed (e.g., Fogassi 2005). Moreover, the IPL has been also associated not only with action understanding, but also with imitation, socialization, empathy, and theory of mind (Gallese 2003, 2007; Dapretto et al. 2005; Oberman et al. 2007). Thus, the ability to simulate behavior of others to be able to prepare a response most beneficial—presumably for the respondent—may have substantially contributed to implement complex processing routines as in the case of linguistic syntax. The creative domain of figurative language may be a phylogenetic relict of the evolving language system available to us today.

References

Aggujaro, S., Crepaldi, D., Pistarini, C., Taricco, M., & Luzzatti, C. (2006). Neuro-anatomical correlates of impaired retrieval of verbs and nouns: Interaction of grammatical class, imageability and actionality. *Journal of Neurolinguistics, 19*(3), 175–194.

Ahrens, K., Liu, H. L., Lee, C. Y., Gong, S. P., Fang, S. Y., & Hsu, Y. Y. (2007). Functional MRI of conventional and anomalous metaphors in Mandarin Chinese. *Brain and Language, 100,* 163–171.

Allport, D. A. (1985). Distributed memory, modular systems and dysphasia. In S. K. Newman & R. Epstein (Eds.), *Current perspective in dysphasia*. Edinburgh: Churchill Livingstone.

Amodio, D. M., & Frith, C. D. (2006). Meeting of minds: The medial frontal cortex and social cognition. *Nature Reviews Neuroscience, 7,* 268–277.

Beauvois, M. F. (1982). Optic aphasia: A process of interaction between vision and language. *Philosophical Transactions of the Royal Society of London. Series B, Biological Sciences, 298*(1089), 35–47.

Beauvois, M. F., Saillant, B., Meininger, V., & Lhermitte, F. (1978). Bilateral tactile aphasia: A tacto-verbal dysfunction. *Brain: A Journal of Neurology, 101*(3), 381–401.

Binder, J. R., Frost, J. A., Hammeke, T. A., Cox, R. W., Rao, S. M., & Prieto, T. (1997). Human brain language areas identified by functional magnetic resonance imaging. *The Journal of Neuroscience, 17*(1), 353–362.

Binder, J. R., Desai, R. H., Graves, W. W., & Conant, L. (2009). Where is the semantic system? A critical review and meta-analysis of 120 functional neuroimaging studies. *Cerebral Cortex, 19*(12), 2767–2796.

Bobrow, S. A., & Bell, S. M. (1973). On catching on to idiomatic expressions. *Memory & Cognition, 1*(3), 343–346.

Bohrn, I. C., Altmann, U., & Jacobs, A. M. (2012). Looking at the brains behind figurative language—a quantitative meta-analysis of neuroimaging studies on metaphor, idiom, and irony processing. *Neuropsychologia, 50*(11), 2669–2683.

Bookheimer, S. (2002). Functional MRI of language: New approaches to understanding the cortical organization of semantic processing. *Annual Review Neuroscience, 25,* 151–188.

Bottini, G., Corcoran, R., Sterzi, R., Paulesu, E., Schenone, P., Scarpa, P., Frackowiak, R. S., & Frith, C. D. (1994). The role of the right hemisphere in the interpretation of figurative aspects of language. A positron emission tomography activation study. *Brain, 117*(6), 1241–1253.

Brownell, H. H., Potter, H. H., Michelow, D., & Gardner, H. (1984). Sensitivity to lexical denotation and connotation in brain-damaged patients: A double dissociation? *Brain and Language, 29,* 310–321.

Brownell, H. H., Simpson, T. L., Bihrle, A. M., Potter, H. H., & Gardner, H. (1990). Appreciation of metaphoric alternative word meanings by left and right brain-damaged patients. *Neuropsychologia, 28*(4), 375–383.

Bruner, E. (2004). Geometric morphometrics and paleoneurology: Brain shape evolution in the genus Homo. *Journal of Human Evolution, 47*(5), 279–303. doi:10.1016/j.jhevol.2004.03.009.

Brunet, E., Sarfati, Y., Hardy-Bayle, M. C. & Decety, J. (2000). A PET investigation of the attribution of intentions with a nonverbal task. *NeuroImage, 11,* 157–166.

Bruner, E. (2010). Morphological differences in the parietal lobes within the human genus: A neurofunctional perspective. *Current Anthropology,* 51(S1), S77–S88.

Buccino, G., Binkofski, F., Fink, G. R., Fadiga, L., Fogassi, L., Gallese, V., Seitz, R. J., Zilles, K., Rizzolatti, G., & Freund, H. J. (2001). Action observation activates premotor and parietal areas in a somatotopic manner: An fMRI study. *European Journal of Neuroscience, 13*(2), 400–404.

Buccino, G., Riggio, L., Melli, G., Binkofski, F., Gallese, V., & Rizzolatti, G. (2005). Listening to action-related sentences modulates the activity of the motor system: a combined TMS and behavioral study. *Brain Research. Cognitive Brain Research, 24*(3), 355–363.

Buxbaum, L. J., Vermonti, T., & Schwartz, M. F. (2000). Function and manipulation tool knowledge in apraxia: Knowing "what for"but not "how". *Neurocase, 6,* 83–97.

Cacciari, C., & Glucksberg, S. (1991). Understanding idiomatic expressions: The contribution of word meanings. In: G. B. Simpson (Ed.), *Understanding word and sentence.* Amsterdam: North-Holland.

Cacciari, C., & Tabossi, P. (1988). The comprehension of idioms. *Journal of Memory and Language, 27,* 668–683.

Calkins, M. W. (1893). A statistical study of pseudo-chromesthesia and of mental-forms. *The American Journal of Psychology, 5*(4), 439–464.

Caplan, D. (2001). Functional neuroimaging studies of syntactic processing. *Journal of Psycholinguistic Research, 30,* 297–320.

Caplan, D., Alpert, N., & Waters, G. (1998). Effects of syntactic structure and propositional number on patterns of regional cerebral blood flow. *Journal of Cognitive Neuroscience, 10,* 541–552.

Caramazza, A., Hillis, A. E., Rapp, B. C., & Romani, C. (1990). The multiple semantics hypothesis: Multiple confusions? *Cognitive Neuropsychology, 7*(3), 161–189.

Caspers, S., Geyer, S., Schleicher, A., Mohlberg, H., Amunts, K., & Zilles, K. (2006). The human inferior parietal cortex: Cytoarchitectonic parcellation and interindividual variability. *NeuroImage, 33*(2), 430–448.

Catani, M., Jones, D. K., & ffytche, D. H. (2005). Perisylvian language networks of the human brain. *Annals of Neurology, 57*(1), 8–16.

Copland, D. A., Chenery, H. J., & Murdoch, B. F. (2002). Hemispheric contributions to lexical ambiguity resolution: Evidence from individuals with complex language impairment following left-hemisphere lesions. *Brain and Language, 81,* 131–143.

Crepaldi, D., Berlingeri, M., & Luzzatti, C. (2013). Clustering the lexicon in the brain: A meta-analysis of the neurofunctional evidence on noun and verb processing. *Frontiers in Human Neuroscience, 7,* 303.

Cronk, B., Lima, S., & Schweigert, W. (1993). Idioms in sentences: Effects of frequency, literalness, and familiarity. *Journal of Psycholinguistic Research, 22*(1), 59–81.

Daniele, A., Giustolisi, L., Silveri, M. C., Colosimo, C., & Gainotti, G. (1994). Evidence for a possible neuroanatomical basis for lexical processing of nouns and verbs. *Neuropsychologia, 32*(11), 1325–1341.

Damasio, A. R., & Tranel, D. (1993). Nouns and verbs are retrieved with differently distributed neural systems. *Proceedings of the National Academy of Sciences, 90*(11), 4957–4960.

Dapretto, M., & Bookheimer, S. Y. (1999). Form and content. *Neuron, 24,* 427–432.

Dapretto, M., Davies, M. S., Pfeifer, J. H., Scott, A. A., Sigman, M., Bookheimer, S. Y., & Iacoboni, M. (2005). Understanding emotions in others: Mirror neuron dysfunction in children with autism spectrum disorders. *Nature Neuroscience, 9*(1), 28–30.

De Renzi, E., & di Pellegrino, G. (1995). Sparing of verbs and preserved, but ineffectual reading in a patient with impaired word production. *Cortex: A Journal Devoted to the Study of the Nervous System and Behavior, 31*(4), 619–636.

Demb, J., Desmond, J., Wagner, A., Vaidya, C., Glover, G., & Gabrieli, J. (1995). Semantic encoding and retrieval in the left inferior prefrontal cortex: A functional MRI study of task difficulty and process specificity. *Journal of Cognitive Neuroscience, 15,* 5870–5878.

Dronkers, N. F., Wilkins, D. P., Van Valin R. D., Jr., Redfern, B. B., & Jaeger, J. J. (2004). Lesion analysis of the brain areas involved in language comprehension. *Cognition, 92*(1–2), 145–177.

Eviatar, Z., & Just, M. A. (2006). Brain correlates of discourse processing: An fMRI investigation of irony and conventional metaphor comprehension. *Neuropsychologia, 44,* 2348–2359.

Farah, M. J., Hammond, K. M., Mehta, Z., & Ratcliff, G. (1989). Category-specificity and modality-specificity in semantic memory. *Neuropsychologia, 27*(2), 193–200.

Fogassi, L. (2005). Parietal lobe: From action organization to intention understanding. *Science, 308*(5722), 662–667.

Foldi, N. S. (1987). Appreciation of pragmatic interpretations of indirect commands. *Brain and Language, 31,* 88–108.

Frattali, C., Hanna, R., McGinty, A. S., Gerber, L., Wesley, R., Grafman, J., & Coelho, C. (2007). Effect of prefrontal cortex damage on resolving lexical ambiguity in text. *Brain and Language, 102,* 99–113.

Friederici, A. D. (2002). Towards a neural basis of auditory sentence processing. *Trends in Cognitive Sciences, 6*(2), 78–84.

Frith, U., & Frith, C. D. (2003). Development and neurophysiology of mentalizing. *Philosophical Transactions of the Royal Society B: Biological Sciences, 358*(1431), 459–473.

Gallagher, H. L., Happé, F., Brunswick, N., Fletcher, P. C., Frith, U., & Frith, C. D. (2000). Reading the mind in cartoons and stories: An fMRI study of 'theory of mind' in verbal and nonverbal tasks. *Neuropsychologia, 38*(1), 11–21.

Gallese, V. (2003). The roots of empathy: The shared manifold hypothesis and the neural basis of intersubjectivity. *Psychopathology, 36*(4), 171–180.

Gallese, V. (2007). Before and below "theory of mind": Embodied simulation and the neural correlates of social cognition. *Philosophical Transactions of the Royal Society B: Biological Sciences, 362*(1480), 659–669.

Gardner, H., & Brownell, H. H. (1986). *Right hemisphere communication battery.* Boston: Psychological Service VAMC.

Gibbs, R. W., Jr. (1980). Spilling the beans on understanding and memory for idioms in conversation. *Memory & Cognition, 8*(2), 149–156.

Gilbert, S. J., Spengler, S., Simons, J. S., Steele, J. D., Lawrie, S. M., Frith, C. D., & Burgess, P. W. (2006). Functional specialization within rostral prefrontal cortex (area 10): A meta-analysis. *Journal of Cognitive Neuroscience, 18,* 932–948.

Gilbert, S. J., Williamson, I. D. M., Dumontheil, I., Simons, J. S., Frith, C. D., & Burgess, P. W. (2007). Distinct regions of medial rostral prefrontal cortex supporting social and non-social functions. *Social Cognitive and Affective Neuroscience, 2,* 217–226.

Giora, R. (1997). Understanding figurative language: The graded salience hypothesis. *Cognitive Linguistics, 7*(1), 183–206.

Gitelman, D. R., Nobre, A. C., Sonty, S., Parrish, T. B., & Mesulam, M. M. (2005). Language network specializations: An analysis with parallel task designs and functional magnetic resonance imaging. *NeuroImage, 26*(4), 975–985.

Goldenberg, G. (2003). Apraxia and beyond: Life and work of Hugo Liepmann. *Cortex, 39*(3), 509–524.

Goodglass, H. (1994). Category-specific lexical dissociations. *Linguistische Berichte. Special Issue: Linguistics and Cognitive Neuroscience, 6,* 49–61.

Grice, H. P. (1975). Logic and conversation. In: P. Cole & J. L. Morgan (Eds.), *Syntax and semantics 3: Speech acts* (pp. 41–58). New York: Academic Press.

Grodzinsky, Y. (2000). The neurology of syntax: Language use without Broca's area. *Behavioral and Brain Science, 23,* 1–71.

Grodzinsky, Y., & Santi, A. (2008). The battle for Broca's region. *Trends in Cognitive Science, 12,* 474–480.

Grossman, M., Peelle, J. E., Smith, E. E., McMillan, C. T., Cook, P., Powers, J., Dreyfuss, M., Bonner, M.F., Richmond, L., Boller, A., Camp, E., and Burkholder, L. (2013). Category-specific semantic memory: Converging evidence from bold fMRI and Alzheimer's disease. *NeuroImage, 68*(0), 263–274.

Gusnard, D. A., Akbudak, E., Shulman, G. L., & Raichle, M. E. (2001). Medial prefrontal cortex and self-referential mental activity: Relation to a default mode of brain function. *Proceedings of the National Academy of Sciences of the United States of America, 98*(7), 4259–4264.

Hart, J., & Gordon, B. (1992). Neural subsystems for object knowledge. *Nature, 359*(6390), 60–64.

Hart, J., Berndt, R. S., & Caramazza, A. (1985). Category-specific naming deficit following cerebral infarction. *Nature, 316*(6027), 439–440.

Hickok, G., & Poeppel, D. (2007). The cortical organization of speech processing. *Nature Reviews Neuroscience, 8*(5), 393–402.

Hillert, D. (1990). *Sprachprozesse und Wissensstrukturen* [German]. Opladen: Westdeutscher.

Hillert, D. (1992). Lexical semantics and aphasia: A state-of-the-art review. *Journal of Neurolinguistics, 7*(1), 1–43.

Hillert, D., & Buračas, G. (2009). The neural substrates of spoken idiom comprehension. *Language and Cognitive Processes, 24*(9), 1370–1391.

Hillert, D., & Swinney, D. (2000). The processing of fixed expressions during sentence comprehension. In A. Cienki, B. J. Luka, M. B. Smith (Eds.), *Conceptual structure, discourse, and language*. Stanford: CSLI.

Hillert, D., Burrington, D. F. H., & Gupta, G. A. (1994). Semantic activation for lexical perception. *Linguistische Berichte. Special issue 6: Linguistics and Cognitive Neuroscience*, 245–268.

Hillert, D. G. (2004). Spared idiom comprehension in aphasia: A single-case approach. *Brain and Language, 89*(1), 207–215.

Hillert, D. G. (2008). On idioms: Cornerstones of a neurological model of language processing. *Journal of Cognitive Science, 9*(2), 193–233.

Hillert, D. G. (2011). Nimm´s nicht so wörtlich [German]. Spektrum: *Gehirn und Geist, 11*, 70–73.

Hubbard, E. M., Arman, A. C., Ramachandran, V. S., & Boynton, G. M. (2005). Individual differences among grapheme-color synesthetes: Brain-behavior correlations. *Neuron, 45*(6), 975–985.

Humphreys, G. W., Riddoch, M. J., & Quinlan, P. T. (1988). Cascade processes in picture identification. *Cognitive Neuropsychology, 5*(1), 67–104.

Humphreys, G. W., Riddoch, M. J., & Price, C. J. (1997). Top-down processes in object identification: Evidence from experimental psychology, neuropsychology and functional anatomy. *Philosophical Transactions of the Royal Society of London. Series B: Biological Sciences, 352*(1358), 1275–1282.

Humphreys, G. W., Hodsoll, J., & Riddoch, M. J. (2009). Fractionating the binding process: neuropsychological evidence from reversed search efficiencies. *Journal of Experimental Psychology. Human Perception and Performance, 35*(3), 627–647.

Husain, M., & Nachev, P. (2007). Space and the parietal cortex. *Trends in Cognitive Sciences, 11*(1), 30–36.

Joanette, Y., Goulet, P., & Hannequin, D. (1990). Right hemisphere and verbal communication. New York: Springer.

Just, M. A., Cherkassky, V. L., Keller, T. A., & Minshew, N. J. (2004). Cortical activation and synchronization during sentence comprehension in high-functioning autism: Evidence of underconnectivity. *Brain, 127*(8), 1811–1821.

Kaan, E., & Swaab, T. Y. (2002). The brain circuitry of syntactic comprehension. *Trends in Cognitive Science, 6,* 350–356.

Kaas, J. H. (2013). The Evolution of brains from early mammals to humans. Wiley Interdisciplinary Reviews. *Cognitive Science, 4* (1), 33–45.

Kempler, D., Van Lancker, D., & Read, S. (1988). Proverb and idiom comprehension in Alzheimer disease. *Alzheimer Disease and Associated Disorders, 2*(1), 38–49.

Kircher, T. T., Brammer, M., Tous Andreu, N., Williams, S. C., & McGuire, P. K. (2001). Engagement of right temporal cortex during processing of linguistic context. *Neuropsychologia, 39*(8), 798–809.

Köhler, W. (1929). *Gestalt psychology*. New York: Liveright.

Kringelbach, M. L. (2005) The orbitofrontal cortex: Linking reward to hedonic experience. *Nature Reviews Neuroscience, 6,* 691–702.

Laurent, J. P., Denhières, G., Passerieux, C., Iakimova, G., Hardy-Baylé, M. C. (2006). On understanding idiomatic language: The salience hypothesis assessed by ERPs. *Brain Research, 1068(1),* 151–160.

Lauro, L. J. R., Tettamanti, M., Cappa, S. F., & Papagno, C. (2008). Idiom comprehension: A prefrontal task? *Cerebral Cortex, 18,* 162–170.

Lee, S. S., & Dapretto, M. (2006). Metaphorical vs. literal word meanings: fMRI evidence against a selective role of the right hemisphere. *NeuroImage, 29*(2), 536–544.

Lee, L., Friston, K., & Horwitz, B. (2006). Large-scale neural models and dynamic causal modelling. *NeuroImage, 30*(4), 1243–1254.

Levelt, W. J., Roelofs, A., & Meyer, A. S. (1999). A theory of lexical access in speech production. *The Behavioral and Brain Sciences, 22*(1), 1–38.

Libon, D. J., Rascovsky, K., Powers, J., Irwin, D. J., Boller, A., Weinberg, D., McMillan, C. T., & Grossman, M. (2013). Comparative semantic profiles in semantic dementia and Alzheimer's disease. *Brain, 136*(8), 2497–2509.

Liepmann, H. (1905). *Ueber Störungen des Handelns bei Gehirnkranken* [German]. Berlin: S. Karger.

Macrae, C. N., Moran, J. M., Heatherton, T. F., Banfield, J. F., & Kelley, W. M. (2004). Medial prefrontal activity predicts memory for self. *Cerebral Cortex, 14,* 647–654.

Mahon, B. Z., & Caramazza, A. (2008). A critical look at the embodied cognition hypothesis and a new proposal for grounding conceptual content. *Journal of Physiology (Paris), 102*(1–3), 59–70.

Marques, J. F., Raposo, A., & Almeida, J. (2013). Structural processing and category-specific deficits. *Cortex, 49*(1), 266–275.

Martin, A., Wiggs, C. L., Ungerleider, L. G., & Haxby, J. V. (1996). Neural correlates of category-specific knowledge. *Nature, 379*(6566), 649–652.

Martin, A., Ungerleider, L. G., & Haxby, J. V. (2000). Category-specificity and the brain: The sensory-motor model of semantic representations of objects. In M. S. Gazzaniga (Ed.), *The new cognitive neurosciences* (pp. 1023–1036). Cambridge: MIT Press.

Mashal, N., Faust, M., & Hendler, T. (2005). The role of the right hemisphere in processing nonsalient metaphorical meanings: Application of principal components analysis to fMRI data. *Neuropsychologia, 43*(14), 2084–2100.

Mashal, N., Faust, M., Hendler, T., & Jung-Beeman, M. (2008). Hemispheric differences in processing the literal interpretation of idioms: Converging evidence from behavioral and fMRI studies. *Cortex, 44*(7), 848–860.

Mason, R. A., & Just, M. A. (2007). Lexical ambiguity in sentence comprehension. *Brain Research, 1146*(18), 115–127.

Mason, R. A., Just, M. A., Keller, T. A., & Carpenter, P. A. (2003). Ambiguity in the brain: What brain imaging reveals about the processing of syntactically implicit sentences. *Journal of Experimental Psychology: Learning, Memory, and Cognition, 29*(6), 1319–1338.

Maurer, D., Pathman, T., & Mondloch, C. J. (2006). The shape of boubas: Sound-shape correspondences in toddlers and adults. *Developmental Science, 9*(3), 316–322.

McCarthy, R., & Warrington, E. K. (1985). Category specificity in an agrammatic patient: The relative impairment of verb retrieval and comprehension. *Neuropsychologia, 23*(6), 709–727.

McCarthy, R. A., & Warrington, E. K. (1988). Evidence for modality-specific meaning systems in the brain. *Nature, 334*(6181), 428–430.

McCarthy, R. A., & Warrington, E. K. (1994). Disorders of semantic memory. *Philosophical Transactions of the Royal Society of London. Series B, Biological Sciences, 346*(1315), 89–96.

McClelland, J. L. & Rogers, T. T. (2003). The parallel distributed processing approach to semantic cognition. *Nature Reviews Neuroscience, 4,* 310–322.
McGlone, M., Glucksberg, S., & Cacciari, C. (1994). Semantic productivity and idiom comprehension. *Discourse Processes, 17,* 167–190.
Miller, E. K. (1999). The prefrontal cortex: Complex neural properties for complex behavior. *Neuron, 22,* 15–17.
Miller, E. K., & Cohen, J. D. (2001). An integrative theory of prefrontal function. *Annual Review of Neuroscience, 24,* 167–202.
Miller, E. K., Freedman, D. J., & Wallis, J. D. (2002). The prefrontal cortex: categories, concepts and cognition. *Philosophical transactions of the Royal Society of London. Series B, Biological Sciences,* 357(1424), 1123–1136.
Nichelli, P., Grafman, J., Pietrini, P., Clark, K., Lee, K. Y., & Miletich, R. (1995). Where the brain appreciates the moral of a story. *Neuroreport, 6*(17), 2309–2313.
Nunn, J. A., Gregory, L. J., Brammer, M., Williams, S. C. R., Parslow, D. M., Morgan, M. J., Morris, R. G., Bullmore, E. T., Baron-Cohen, S., & Gray, J. A. (2002). Functional magnetic resonance imaging of synesthesia: Activation of V4/V8 by spoken words. *Nature Neuroscience, 5*(4), 371–375.
Oberman, L. M., Pineda, J. A., & Ramachandran, V. S. (2007). The human mirror neuron system: A link between action observation and social skills. *Social Cognitive and Affective Neuroscience, 2*(1), 62–66.
Oliveri, M., Finocchiaro, C., Shapiro, K., Gangitano, M., Caramazza, A., & Pascual-Leone, A. (2004a). All talk and no action: A transcranial magnetic stimulation study of motor cortex activation during action word production. *Journal of Cognitive Neuroscience, 16*(3), 374–381.
Oliveri, M., Romero, L., & Papagno, C. (2004b). Left but not right temporal involvement in opaque idiom comprehension: A repetitive transcranial magnetic stimulation study. *Journal of Cognitive Neuroscience, 16,* 848–855.
Orban, G. A., Claeys, K., Nelissen, K., Smans, R., Sunaert, S., Todd, J. T., Wardak, C., Durand, J. B., & Vanduffel, W. (2006). Mapping the parietal cortex of human and non-human primates. *Neuropsychologia, 44*(13), 2647–2667.
Ortony, A., Schallert, D. L., Reynolds, R. E., & Antos, S. J. (1978). Interpreting metaphors and idioms: Some effects of context on comprehension. *Journal of Verbal Learning and Verbal Behavior, 17,* 465–477.
Papagno, C. & Caporali, A. (2007). Testing idiom comprehension in aphasic patients: The effects of task and idiom type. *Brain and Language, 100*(2), 208–220.
Papagno, C., Oliveri, M. & Romero, L. (2002). Neural correlates of idiom comprehension. *Cortex, 38,* 895–898.
Papagno, C., Curti R., Rizzo, S., Crippa, F., & Colombo, M. R. (2006). Is the right hemisphere involved in idiom comprehension? *A neuropsychological study. Neuropsychology, 20*(5), 598–606.
Patterson, K., Nestor, P. J., & Rogers, T. T. (2007). Where do you know what you know? The representation of semantic knowledge in the human brain. *Nature Reviews Neuroscience, 8*(12), 976–987.
Paul, L., Van Lancker-Sidtis, D. R., Schieffer, B., Dietrich, R., & Brown, W. (2003). Communicative deficits in agenesis of the corpus callosum: Nonliteral language and affective prosody. *Brain and Language, 85,* 313–324.
Perani, D., Cappa, S. F., Bettinardi, V., Bressi, S., Gorno-Tempini, M., Matarrese, M., & Fazio, F. (1995). Different neural systems for the recognition of animals and man-made tools. *Neuroreport, 6*(12), 1637–1641.
Peterson, R. R. & Burgess, C. (1993). Syntactic and semantic processing during idiom comprehension: Neurolinguistic and psycholinguistic dissociation. In: C. Cacciari & P. Tabossi (Eds.), *Idioms: Processing, structure and interpretation.* Hillsdale: Lawrence Erlbaum.
Peterson, R. R., Burgess, C., Dell, G. S., & Eberhard, K. M. (2001). Dissociation between syntactic and semantic processing during idiom comprehension. *Journal of Experimental Psychology: Learning, Memory and Cognition, 27*(5), 1223–1237.

Pineda, J. A. (Ed.). (2009). *Mirror neuron systems—The role of mirroring processes in social cognition*. New York, NY: Humana Press.

Poldrack, R. A., Wagner, A. D., Prull, M. W., Desmond, J. E., Glover, G. H., & Gabrieli, J. D. (1999). Functional specialization for semantic and phonological processing in the left inferior prefrontal cortex. *Neuroimage, 10*(1), 15–35.

Pulvermüller, F., Hauk, O., Nikulin, V. V., & Ilmoniemi, R. J. (2005). Functional links between motor and language systems. *The European Journal of Neuroscience, 21*(3), 793–797.

Ramachandran, V. S., & Hubbard, E. M. (2001). Synaesthesia—A window into perception, thought and language. *Journal of Consciousness Studies, 8*(12), 3–34.

Rapp, A. M., Leube, D. T., Erb, M., Grodd, W., & Kircher, T. T. (2004). Neural correlates of metaphor processing. *Brain Research. Cognitive Brain Research, 20,* 395–402.

Rapp, A. M., Mutschler, D. E., & Erb, M. (2012). Where in the brain is nonliteral language? A coordinate-based meta-analysis of functional magnetic resonance imaging studies. *NeuroImage, 63*(1), 600–610.

Riddoch, M. J., Humphreys, G. W., Coltheart, M., & Funnell, E. (1988). Semantic systems or system? Neuropsychological evidence re-examined. *Cognitive Neuropsychology, 5,* 3–25.

Rizzolatti, G., & Craighero, L. (2004). The mirror-neuron system. *Annual Review of Neuroscience, 27,* 169–192.

Roskies, A. L., Fiez, J. A., Balota, D. A., Raichle, M. E., & Petersen, S. E. (2001). Task-dependent modulation of regions in the left inferior frontal cortex during semantic processing. *Journal of Cognitive Neuroscience, 13*(6), 829–843.

Sahin, N. T., Pinker, S., Cash, S. S., Schomer, D., & Halgren, E. (2009). Sequential processing of lexical, grammatical, and phonological information within Broca's area. *Science, 326*(5951), 445–449.

Shallice, T. (1987). Impairments of semantic processing: Multiple dissociations. In M. Coltheart, G. Sartori, & R. Job (Eds.), *The cognitive neuropsychology of language*. Hillsdale: Lawrence Erlbaum.

Shibata, M., Toyomura, A., Itoh, H., & Abe, J. (2010). Neural substrates of irony comprehension: A functional MRI study. *Brain Research, 1308,* 114–123.

Sotillo, M., Carretie, L., Hinojosa, J. A., Tapia, M., Mercado, F., Lopez-Martin, S., & Albert, J. (2005). Neural activity associated with metaphor comprehension: Spatial analysis. *Neuroscience Letters, 373*(1), 5–9.

Sperber, D., & Wilson, D. (2001). Précis of relevance: Communication and cognition. *Behavioral and Brain Sciences, 10*(4), 697–754.

Stowe, L. A., Broere, C. A. J., Paans, A. M. J., Wijers, A. A., Mulder, G., Vaalburg, W., & Zwarts, F. (1998). Localizing components of a complex task: Sentence processing and working memory. *Neuroreport, 9*(13), 2995–2999.

Stringaris, A. K., Medford, N. C., Giampietro, V., Brammer, M. J., & David, A. S. (2007). Deriving meaning: Distinct neural mechanisms for metaphoric, literal, and non-meaningful sentences. *Brain and Language, 100,* 150–162.

Subramaniam, K., Beeman, M., & Mashal, N. (2013). Positively valenced stimuli facilitate creative novel metaphoric processes by enhancing medial prefrontal cortical activation. *Frontiers in Cognitive Science, 4,* 211.

Swinney, D., & Cutler, A. (1979). The access and processing of idiomatic expressions. *Journal of Verbal Learning and Verbal Behavior, 18,* 523–534.

Tabossi, P., & Zardon, F. (1993). The activation of idiomatic meaning in spoken language comprehension. In C. Cacciari & P. Tabossi (Eds.), *Idioms: Processing, structure and interpretation*. Hillsdale: Lawrence Erlbaum.

Tabossi, P., & Zardon, F. (1995). The activation of idiomatic meaning. In M. Everaert, E. J. van der Linden, A. Schenk, & R. Schreuder (Eds.), *Idioms: structural and psychological perspectives*. Hillsdale: Lawrence Erlbaum.

Titone, D. A. & Connine, C. M. (1994). Comprehension of idiomatic expressions: Effects of predictability and literality. *Journal of Experimental Psychology: Learning, Memory, and Cognition, 20*(5), 1126–1138.

Tompkins, C. A., Boada, R., & McGarry, K. (1992). The access and processing of familiar idioms by brain-damaged and normally aging adults. *Journal of Speech, Language and Hearing Research, 35*(3), 626–637.

Turken, A. U., & Dronkers, N. F. (2011). The neural architecture of the language comprehension network: Converging evidence from lesion and connectivity analyses. *Frontiers in Systems Neuroscience, 5,* 1–20.

Uchiyama, H., Seki, A., Kageyama, H., Saito, D. N., Koeda, T., Ohno, K., & Sadato, N. (2006). Neural substrates of sarcasm: A functional magnetic-resonance imaging study. *Brain Research, 1124*(1), 100–110.

Van Lancker, D. R., & Kempler, D. (1987). Comprehension of familiar phrases by left- but not by right-hemisphere damaged patients. *Brain and Language, 32,* 265–277.

Van Lancker-Sidtis, D. (2004). When novel sentences spoken or heard for the first time in the history of the universe are not enough: Toward a dual-process model of language. *International Journal of Language and Communicative Disorders, 39*(1), 1–44.

Wagner, A., Desmond, J., Demb, J., Glover, G., & Gabrieli, J. D. (1997). Semantic repetition priming for verbal and pictorial knowledge. *Journal of Cognitive Neuroscience, 9,* 714–726.

Wakusawa, K., Sugiura, M., Sassa, Y., Jeong, H., Horie, K., Sato, S., Yokoyama, H., Tsuchiya, S., Inuma, K., & Kawashima, R. (2007). Comprehension of implicit meanings in social situations involving irony: A functional MRI study. *Neuroimage, 37*(4), 1417–1426.

Warrington, E. K., & Shallice, T. (1984). Category specific semantic impairments. *Brain, 107*(3), 829–854.

Winner, E. (1988). *The point of words: Children's understanding of metaphor and irony*. Cambridge: Harvard University Press.

Winner, E. & Gardner, H. (1977). The comprehension of metaphor in brain-damaged patients. *Brain, 100,* 719–727.

Worsley, K. J., Evans, A. C., Marrett, S., & Neelin, P. (1992). A three-dimensional statistical analysis for CBF activation studies in human brain. *Journal of Cerebral Blood Flow and Metabolism, 12,* 900–918.

Zempleni, M. Z., Haverkort, M., Renken, R., & Stowe, L. A. (2007a). Evidence for bilateral involvement in idiom comprehension: An fMRI study. *Neuroimage, 34*(3), 1280–1291.

Zempleni, M. Z., Renken, R., Hoeks, C. J., Hoogduin, J. M., & Stowe, L. A. (2007b). Semantic ambiguity processing in sentence context: Evidence from event-related fMRI. *Neuroimage, 34,* 1270–1279.

Chapter 11
Atypical Language

11.1 Aphasia

One of the most important approaches in the history of cognitive neuroscience refers to the meticulous analysis of language disorders in particular clinical populations or in single cases. The medical philosopher, Alcmaeon of Croton (ca. BC 570–490) and the physician, Galen von Pergamon (also known as Claudius Galenus: AD 129 – ca. 217 or earlier) were one of the first, who claimed that the brain (and not the heart as believed by Aristotle: BC 384–322) is the organ that generates mental activity. Other significant milestones were the concept of phrenology (Franz Joseph Gall: 1758–1828), the discoveries of Broca's area and Wernicke's area respectively by the French physician, Paul Pierre Broca (1824–1880) and the German physician Carl Wernicke (1848–1904), the function of neurons by Santiago Ramón y Cajal (1852–1932) and Camillo Golgi (1843–1926), or the mathematical groundwork for neuroimaging by Allan Cormack (1924–1998), and Godfrey Hounsfield (1919–2004). A review of the history provides an exciting awareness of the scientific development, in general and in particular, on the neurobiological foundations of cognition.

A wide range of different clinical disorders are associated with language disorders. The risk of developing specific language disorders such as aphasia or general cognitive impairments along with language disorders such as in Alzheimer's disease is age related. However, neurological injuries or diseases that cause language disorders can occur at any stage during a lifecycle of a person. Stroke-related aphasic syndromes reveal language disorders at different linguistic levels, which are relatively systematic, and subjects with autism often show specific impairments at the interface of language and cognition, at the pragmatic level. Despite the application of modern neuroimaging techniques, often the behavioral analyses of single clinical cases provide surprising results, which would not have been discovered otherwise. This clinical approach is not only of theoretical interest as any new knowledge and findings provide information for better treatment procedures of these specific speech and language disorders.

According to the National Stroke Association in 2008, there are 80,000 new cases of aphasia in USA and the National Institute of Neurological Disorders and

D. Hillert, *The Nature of Language*, DOI 10.1007/978-1-4939-0609-3_11,

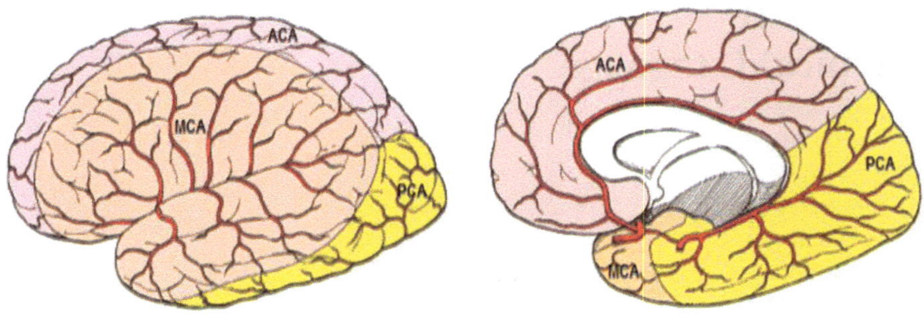

Fig. 11.1 Sagittal (*left*) and mid-sagittal (*right*) diagram of the human brain: blood supply to regions by three main arteries (*ACA* anterior cerebral artery, *MCA* middle cerebral artery, *PCA* posterior cerebral artery. (Adapted and modified, © James Publishing)

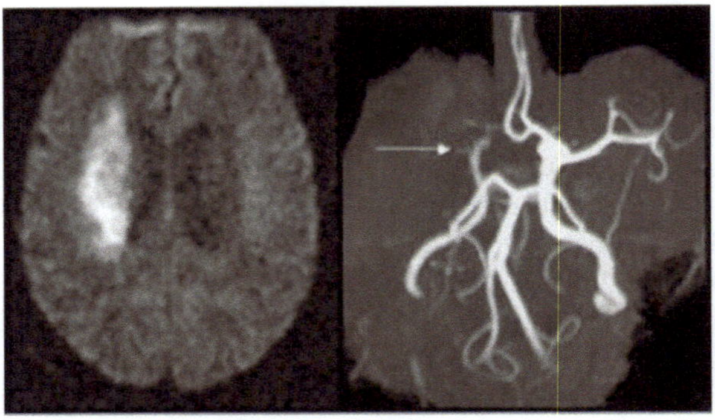

Fig. 11.2 Diffusion-weighted imaging shows a cerebral infarct (bright area). Without blood flow as in the case of a stroke, the diffusion (molecular movement of water) is impaired in the tissues (left side). Magnetic resonance angiography reveals an acute stroke with occlusion of the middle cerebral artery (right side). (Adapted and modified; © University Medical Center, Freiburg)

Stroke estimates a prevalence of approximately 1 million patients (1 in 250). It is projected that by 2020, the incidence rate would be 180,000/year and prevalence of 2 million patients with aphasia. The primary cause includes strokes (85 %), other causes are traumatic injuries, tumors, or medical procedures. Most strokes are caused by a cerebral infarction, that is, by an arterial thrombosis (blood clot). Typically, a blood clot forms around fatty plaques. Cerebral arteries cannot be cleaned out of fatty blockages unlike arteries in the heart or leg. The other two major causes are cerebral hemorrhage and cerebral aneurysm. In the case of a hemorrhage, uncontrolled hypertension often leads to the rupture of the cerebral artery. A cerebral aneurysm is typically caused by the rupture of weak and enlarged cerebral artery (see Figs. 11.1 and 11.2).

Aphasia research studies prefer to focus on the examination of Broca's and Wernicke's aphasia as they affect all language-specific levels from phonology and syntax to semantics (e.g., Caplan 1987; Hillert 1990; Peng 2009; Ingram 2007).

Broca's aphasic patients suffer from a severe language production disorder, which is typically caused by stroke lesions to BA 44 of the left inferior frontal gyrus (IFG). The speech consists of short phrases, often produced with effort or haltingly. This is the most notable feature of Broca's aphasia (Pick 1913). This syndrome is not the result of pure motor inferences with the execution of fluent speech, because the omission of words is selective. Typically, the production of Broca's aphasic patients consists of nouns and verbs, but grammatical words (closed-class words) are omitted. The classical connectionist schools believed in cerebral language centers and the emphasis was on the performance in different verbal modalities (Broca 1861; Wernicke 1874; Kussmaul 1881; Lichtheim 1884).

Since the 1960th, psycholinguistic research challenged the view of the classical language centers (e.g., Goodglass and Berko 1960; Caramazza and Zurif 1976). The functions of the language centers are not regarded to reflect linguistic processes *per se* but systems for the analysis and synthesis of language. While Broca's area was regarded to generate syntax (algorithmic analysis), Wernicke's area was characterized as generating meanings (heuristic analysis). Syntactic deficits in Broca's aphasia were found in production and comprehension. However, several individuals showed only an expressive agrammatism while syntactic comprehension was spared (e.g., Nespoulous et al. 1988; Miceli et al. 1983), and some patients with receptive agrammatism appeared not to have expressive agrammatism at all (Zurif and Caramazza 1976; Martin 1987).

In the 1980s, more precise models were developed to account for agrammatic disorders. One account claimed that a specific lexical route was impaired (Bradley 1978; but see Gordon and Caramazza 1983; Kolk and Blomert 1985); agrammatic patients have lost the ability to access grammatical elements, which results in a syntactic deficit. This account holds that syntactic representations are spared while access to grammatical words is deficient. Another account, the syntactic mapping hypothesis, postulated that agrammatic Broca's aphasic patients are not able to map syntactic functions onto matching θ-roles while their ability to judge grammatical information is not affected (Schwartz et al. 1987). Finally, it has been claimed that Broca's aphasic patients fail to produce and comprehend suffixes and closed-class words (and thus are not able to parse sentences), because they do not exist autonomously as phonological words (Kean 1977). Independent of these different accounts, which all certainly address important viewpoints, it is obvious that Broca's aphasic patients have difficulties to parse only specific types of syntactic structures.

Typically, Broca's aphasic patients are able to compute canonical syntactic structures by assigning θ-roles to particular syntactic categories. However, they have difficulties with some (but not all) intrasentential dependencies. Moreover, individual patients show different patterns of morpheme retention. For example, Nespoulous et al. (1988) described a French speaking Broca's aphasic patient, who failed to produce auxiliaries and weak forms of pronouns such as *le, la, and lui*, but was able to produce strong forms such as *elle, il, moi, and toi*. Again, Miceli et al. (1983) described an Italian speaking patient, who could not produce verb inflections, but had no difficulties with closed-class elements. In contrast, Goodglass' (1976) English speaking patient was able to inflect verbs. Serbo-Croatian speaking patients seem to be able to use Case markers on noun phrases (Lukatela et al. 1988). This large variability indicates that closed-class items do not belong to a unitary category. An alternative account argues that it is a temporal problem rather than a structural one. Due to temporal delays, Broca's aphasic patients cannot access syntactic information in time, although their grammatical knowledge seems to be spared (Swinney et al. 1989; Zurif et al. 1993). However, a temporal processing disorder does not account for selective impairments of morpheme structures across languages as well as within a single language.

In the past, most studies could not find evidence for semantic deficits in Broca's aphasia (e.g., Blumstein et al. 1982; Katz 1988). However, some studies reported delayed or deficient lexical–semantic processing due to an increase of computational costs (Prather et al. 1997; Swinney et al. 1989; Milberg and Blumstein 1981; Milberg et al. 1987). Several neuroimaging studies show that BA 44 is not exclusively related to Broca's aphasia (e.g., Mohr 1976; Alexander et al. 1990; Dronkers 1996), and that the left IFG does not only involve phonological or syntactic processing, but also lexical–semantic computations (e.g., Poldrack et al. 1999; see Bookheimer 2002 for a review). Particularly, BA 47, which is located between the pars triangularis and pars orbitalis, appears to play a crucial role in manipulating or generating lexical–semantic information. However, to what extent the left-IFG contributes to semantic processing is still controversial (e.g., Thompson-Schill et al. 1999; Wagner et al. 2001; Thompson-Schill 2003). While it has been shown that in some cases the right cortex is also involved in spontaneous recovery from aphasic deficits (e.g., Musso et al. 1999; Thompson 2000; Hillis 2002), it less clear how much intensive lexical training can improve the linguistic recovery process in aphasia as the effects are usually based on multiple factors and specific to the linguistic neuro wiring of an individual patient.

In contrast, Wernicke's aphasic patients suffer from severe language comprehension deficits that involve left-hemisphere lesions to the SMG (BA 40) and AG (BA 39) in addition to the STG (BA 22) and the MTG (BAs 21 & 37) (e.g., Kertesz et al. 1993). The common definition of Wernicke's area includes the posterior third of the STG. However, a lesion restricted to this area does not give rise to the long-lasting symptom complex, termed Wernicke's aphasia (Selnes et al. 1983). Also, patients have been reported who have lesions outside of Wernicke's area (SMG and AG) and who show the symptoms of Wernicke's aphasia (e.g., Murdoch et al. 1986; Damasio et al. 1996). Lesion data reveal left, middle, and inferior temporal lobe activations during semantic processing, which has been confirmed by neuroimaging studies (e.g., Vandenberghe et al. 1996, Binder et al. 1997). Surprisingly, fMRI experiments also revealed activations in the left inferior frontal gyrus when participants are asked to categorize, generate, or judge lexical–semantic information (see Poldrack et al. 1999 for a review). The left IFG is the typical lesions site of Broca's aphasia. However, the status of lexical processing is quite controversial. It is not premature to claim that both, the left temporo-parietal region as well as the left frontal region, are involved in lexical processing. One assumption is that lexical representations are stored implicit in left posterior temporo-parietal regions and executing these lexical representations involves left anterior language regions such as the IFG (Roskies et al. 2001; Bookheimer 2002).

To what extent the right hemisphere is typically involved in lexical processing is less clear. However, investigating the role of the linguistically nondominant right hemisphere in right handers became particularily popular by split-brain studies (Gazzaniga and Sperry 1967; Baynes et al. 1992; Gazzaniga 1995). Their findings as well as results from right-hemispheric lesion studies in right handers indicate that the right hemisphere processes lexical information at the word level, but not syntactic and phonological information (e.g., Gainotti et al. 1981). Moreover, lesion studies revealed idiomatic or metaphoric processing deficits in right-lesioned patients (e.g., Van Lancker and Kempler 1987; Molloy et al. 1990; Van Lancker 1990). For instance, in Van Lancker and Kempler's (1987) study, right-lesioned patients showed impairments when asked to select pictured representations of sentences containing familiar phrases, but not when asked to choose literal sentence meanings. In opposition, aphasic patients performed more poorly than the right-lesioned group in (offline) sentence comprehension tasks (e.g., Winner and Gardner 1977; Myers and Linebaugh 1981; Tompkins et al. 1992). However, Tompkins et al. (1992) found with a word-monitoring task, which is considered an online task, that not only right-lesioned patients, but also aphasic patients access idiomatic meanings (see also Hillert 2004).

As previously mentioned, Bottini et al. (1994) conducted the first PET (positron emission tomography) study on figurative processing to examine plausibility judgments of metaphoric and literal phrases. They reported relatively greater activation in different regions of the right hemisphere, in particular, in the IFG, the premotor cortex, and the posterior temporal lobe. These right-hemispheric regions are roughly homologs of (the left-sided) Broca's and Wernicke's areas. Bottini and colleagues conclude that in contrast to processing literal meanings, judging metaphors require reference to long-term, episodic memories and they assume that frontal lobe activations may reflect the search for long-term memories or the generation of visual imagery to facilitate decision making. Nichelli et al.'s (1995) PET study confirmed to some extent Bollini et al.'s findings. In Nichelli et al.'s study, participants produced judgments about the metaphoric or literal meaning of fables. In addition to left-hemispheric activations, figurative judgments (e.g., the moral of a story) produced a relative increase of activity compared to literal details (e.g., semantic or syntactic judgments). However, the right-sided activation might not be necessarily due to figurative inferences, but due to the process of drawing inferences *per se* or usage of pragmatic context in judging the passage.

PET studies, performed months to years after recovery, reveal spontaneous reorganization of functions to the right hemisphere (e.g., Naeser and Palumbo 1994; Weiller et al. 1995; Ohyama et al. 1996; Musso et al. 1999; Warburton 1999; Heiss et al. 1999). Thus, even the adult brain has mechanisms of plasticity that can produce rapid rehabilitation. Functional neuroimaging is important to distinguish the success of stroke interventions from innate compensatory mechanism. To achieve a higher level of language proficiency, reorganization of a net of cortical areas, specialized in one or more aspects of language processing, is necessary (e.g., Mesulam

1994; Thompson 2000). When a key node of a large-scale cortical net is damaged by a stroke, undamaged net components (i.e., contralateral homologs) are increasingly activated. Since the workload of the remaining net is modified, a shift in cognitive workload can occur toward the contra-lateral hemisphere (Thulborn et al. 1999; Frackowiak 2001). Interestingly, with increasing age, no decreasing contribution of the right hemisphere to language functions occurs (Nocentini et al. 1999).

In studying the linguistic, and in particular lexical disorders, it is important to take into account a psycholinguistic approach that considers to some extent a theoretical account of cognitive processing. The distinction between lexical perception and lexical composition is here of particular relevance (e.g., Forster 1979; Fodor 1983; Seidenberg 1985). Lexical perception takes place automatically and relatively fast and without conscious control. They are therefore regarded to be impenetrable, static and largely bottom-up (data-) driven. In contrast, lexical composition involves controlled conscious processes, which are dynamic, flexible, effortful, relatively slow in comparison to automatic computations, and knowledge driven. While offline tasks seem to test controlled compositional processes, online tasks examine perceptual automatic processes (Posner and Snyder 1975; Schneider and Shiffrin 1977; Shiffrin and Schneider 1977). Both types of processes seem to be independent of each other when automated processes have been stabilized *per se*.

Typically, Wernicke's aphasic patients have difficulties in processing content words and Broca's aphasic patients appear to be impaired in processing clitics like bounded morphemes (affixes) or function words. Numerous offline studies seem to support the so-called semantic deficit hypothesis not only in Wernicke's aphasia but also to some degree in Broca's aphasia (e.g., Lesser 1978; Goodglass and Baker 1976; Caramazza et al. 1982). However, the status of semantic processing in aphasia is quite unclear, because the response patterns in offline studies seem to reflect the product of possibly different compensatory strategies in Broca's and Wernicke's aphasia. Also, a series of online tasks, that examined (automatic) lexical–semantic priming, indicates that Wernicke's aphasic patients' access to (literal) lexical meanings is spared. Broca's aphasic patients, however, seem to be sensitive to lexical–semantic priming in some experiments (e.g., Blumstein et al. 1982; Katz 1988; Ostrin and Tyler 1993), and in other studies they did not show significant priming (e.g., Milberg and Blumstein 1981; Milberg et al. 1987; Swinney et al. 1989). The factor that appears to determine Broca's aphasics' performance is obviously dependent on the computational costs involved in the task condition (e.g., word triplets vs. word pairs, or primary meaning vs. secondary meaning). These priming patterns are therefore compatible with recent findings that the left IFG plays a crucial role in lexical processing. Cumulative evidence suggests that lexical processing *per se* is accomplished by large scale, variable, and distributed patterns of activities among interactive cortical areas. Single processes are associated with activations in multiple regions and the site of the lesion is poorly correlated with the behavioral deficits (Caplan and Hildebrandt 1988). These findings are more compatible with interactive models of language processing that stress the mutual influence and cascaded interaction of linguistic components (McClelland 1987).

The use of neuroimaging techniques for examining post-stroke recovery processes certainly plays today a very important role in the treatment program. One fMRI study revealed that during an acute stage (mean 1.8 days post-stroke), depressed overall brain activation was found, in particular left sided (Saur et al. 2006). During the recovery stage (mean 12 days post-stroke), overall increase of brain activation was measured, but this time particularly strong in the right hemisphere. Finally, about 320 days post-stroke, the left hemisphere showed the strongest activation and correlated with aphasia recovery. In remains unclear what kind of early interventions during which recovery stage might have a positive effect on cortical reorganization. It appears in general that recovery improves along with a decrease of right-hemisphere activities (in typical right-handed aphasic patients) as spared cortical structures of the left hemisphere take over damaged left-sided structures (Cornelissen et al. 2003; Meinzer et al. 2008; Postman-Caucheteux et al. 2010; Rosen et al. 2000). In Schlaug et al.'s (2009) diffusion tensor imaging (DTI) study, the effect of melodic intonation therapy on white-matter density in the right hemisphere was examined in nonfluent aphasic patients (global or Broca's aphasia). The idea of a melodic intonation therapy is that singing might improve the patients' speech. After completion of the therapy, patients had increased white-matter fibers in the right hemisphere. Thus, behavioral interventions can target specific cortical areas to improve communication skills in aphasia. However, behavioral interventions must be individually adjusted. For instance, in the case of small left cortical lesions, a right-sided focal intervention may be less appropriate as compared to a left-sided focal intervention. Also, attempts to stimulate specific spared cortical areas, by applying for instance transcranial magnetic stimulation (TMS), require the use of MRI to precisely localize the damaged area to be excluded from procedure. A better understanding of the neural processes underlying spontaneous recovery is significant for providing effective treatment of communicative disorders in aphasia.

11.2 Communicative Disorders

One known population of patients, which is at relatively high risk of developing Alzheimer's disease (AD), suffer from mild cognitive impairment (MCI). According to the Peripheral and Central Nervous System Drugs Advisory Committee, the conversion rate from MCI to AD is 80% in 10 years or 10–15% per year. In contrast, the conversion rate for healthy elderly subjects to MCI or AD is 1–2% per year. MCI might therefore be considered as an early stage of AD rather than as a distinct condition (Morris et al. 2001). It is unclear, how reliably specific symptoms of MCI do indeed reflect early stages of AD. Most of the research on MCI has been done on the amnesic form. However, as the concept of an intermediate state of impairment between normal aging and fully developed AD has evolved, it has become apparent that there is more heterogeneity to MCI (Lopez et al. 2003; Petersen 2003). For example, individuals may be impaired in a cognitive domain other than memory (e.g., language or executive functions). Alternatively, an MCI patient may have memory problems in addition to difficulties in complex language processing and executive functions, although insufficiently severe to be diagnosed with AD. Therefore, the clinical concept of MCI includes at least three subtypes: amnestic, multidomain, and single nonamnestic domain MCI. In Lopez et al.'s (2003) cardiovascular health study cognition study, MCI (22 % prevalence at the Pittsburgh Center) was associated with race (African American), low educational level, low MMSE (Mini-Mental State Examination) scores and Digit Symbol Test (DST), cortical atrophy, MRI identified infarcts, and measurements of depression. Amnestic MCI (6 % prevalence) was associated with MRI identified infarcts, the presence of the APOE 4 (apolipoprotein E) allele[1], and low (modified) MMSE scores. Multidomain MCI (16 % prevalence) was associated with low modified MMSE and DST scores. Amnestic MCI is at high risk of developing AD rather than other forms of dementia. Multidomain MCI patients may develop AD, but may be also a phenotype of incipient vascular dementia. Single nonamnestic domain MCI is often the harbinger of non-AD dementias (e.g., frontotemporal dementia or Lewy body dementia) rather than AD.

While there is no doubt that AD patients' language processing abilities are to some extent reduced or impaired, it is quite controversial how to interpret these deficits to provide clinical recommendations. Earlier findings seem to indicate that syntax is spared in AD compared to lexical semantics (e.g., Appell et al. 1982; Bayles 1982; Cummings et al. 1985; Murdoch et al. 1987). In particular, naming impairments (e.g., Huff et al. 1986; Bayles and Tomoeda 1983; Schwartz et al. 1979; Martin and Fedio 1983), lexical comprehension, and fluency disorders (Martin and Fedio 1983; Kertesz et al. 1986) have been considered as evidence for impaired lexical semantics along with apparently preserved syntax in AD. In addition, this view was also supported by the observation that (at least mild) AD patients are able to produce relatively complex sentence structures (e.g., Hier et al. 1985; Kemper et al. 1993).

However, most studies that support the view of impaired lexical semantics and spared syntax referred to the behavior of AD patients in language production tasks. Specific language studies revealed that AD patients also have sentence comprehension difficulties (e.g., Emery and Breslau 1989, Kemper et al. 1993; Grossman et al. 1995, 1996, 1998; Rochon et al. 1994; Grossman and White-Devine 1998; Almor et al. 1999; Bickel et al. 2000). For example, that AD patients do not understand sentences as age-matched control subjects do, has been demonstrated in a variety of offline (nontime constrained) tasks such as sentence–picture matching, enactment,

[1] The apolipoprotein E (APOE) gene (chromosome 19) is the main genetic cause for late-onset AD. This gene has three allelic variants (2, 3, and 4) and five common genotypes (2/3, 3/3, 2/4, 3/4, and 4/4). The APOE-4 allele increases the risk and decreases the average age of dementia. The risk of AD is lowest in patients with the 3/3 genotype, higher for the 3/4 genotype, and highest for the 4/4 genotype (Corder et al. 1994).

and the Token Test[2]. It is unclear which specific cognitive or neurological deficits cause these sentence comprehension disorders. A combination of different linguistic (e.g., processing of conceptual, semantic, thematic, and syntactic information) and/or more general cognitive factors (e.g., general-purpose WM or specialized WM functions) may all contribute to sentence processing difficulties in AD.

However, lexical–semantic processing deficits are the most characteristic disorders found in AD patients (Nebes et al. 1989). Their semantic memory is severely impaired, which is evident in their spontaneous speech as well as in their performance on tests of verbal fluency and object naming (e.g., Grober et al. 1985; Huff et al. 1986; Butters et al. 1987). Two main accounts were discussed in the literature to predict the performance of AD in semantic processing task. One hypothesis claims a procedural deficit of semantic information, while the second hypothesis postulates a degraded semantic storage deficit (e.g., Hodges et al. 1992; Cross et al. 2008). Another approach considers the distinction between automatic vs. controlled semantic processing (Shiffrin and Schneider 1977). Some studies reported preserved semantic priming in AD, some no priming and some hyperpriming (e.g., Ober and Shenaut 1988; Salmon et al. 1988; Chertkow et al. 1989; Nebes et al. 1986; Balota and Duchek 1991).

However, there are several reasons why lexical–semantic deficits alone do not explain AD patients' sentence comprehension difficulties. One reason is that AD patients have word finding difficulties in object naming, but understanding of single words is relatively spared. Another reason is that AD patients have sentence comprehension difficulties despite the fact of relatively intact lexical comprehension. Thus, one may conclude that AD patients' sentence processing difficulties are not related to their naming or semantic processing disorders that have been reported at the word level. Some investigators believe therefore that deficits at the sentence level are caused by syntactic deficits or reduced memory capacities in AD rather than by lexical–semantic deficits. However, in considering the fact that AD patients show lexical–semantic disorders in a variety of different task conditions (Bayles et al. 1990; Chertkow and Bub 1990), the hypothesis that sentence comprehension deficits results from lexical–semantic deficits cannot be excluded. In particular, the idea that AD patients' sentence processing difficulties increase with the number of propositions a sentence carries might partly be caused by semantic impairments (Rochon et al. 1994). In addition, a deficit in comprehending verbs (and the thematic structures provided by the verb) might be also compatible with the number of proposition hypothesis (Grossman et al. 1996; Grossman and White-Devine 1998). An alternative explanation is that the number of proposition hypothesis is related to a deficit of postinterpretive processing (Rochon et al. 1994; Waters et al. 1998). This account is also compatible with the view that AD patients suffer from poor inhibitory control or metalinguistic process difficulties (Hillert 1999). For example, Kempler et al. (1998) used a cross-modal naming task to examine how AD patients comprehend violations of subject-verb agreement and verb transitivity. The

[2] The Token Test, originally developed by De Renzi and Vignolo (1962), uses different types of tokens (colored squares and circles in small and large sizes) to assess language comprehension.

patients, who failed in sentence–picture matching also showed a delay in naming the target word when a sentence fragment involved a grammatical error.

Almor et al. (1999) argue however that a reduced WM capacity causes less sensitivity to the appropriate pronoun forms compared to the matching noun phrases. Grossman and Rhee (2001) applied a word-monitoring task to examine sensitivity to grammatical and semantic agreement violations. In contrast to control subjects, AD patients did not show a significant delay in their response to a target word in the immediate vicinity of an agreement violation. These data seem to indicate an abnormal time course of sentence processing in AD patients. Again, in using an auditory moving-window paradigm, Waters and Caplan (2002) found that reduced WM functions in AD were not correlated to their ability to access semantic and syntactic information while processing phrase-by-phrase different sentence types. They concluded that online (time-constrained) syntactic processing is not impaired in AD patients despite their reduced WM capacities. Instead, Waters and Caplan discuss the possibility of a specialized WM system for online, first-pass interpretative processing and one for postinterpretative processing. They conclude that spared syntactic processing in AD as well as severe (general) WM deficits point to a deficit of postinterpretative processing in AD.

The degradation of brain structures in AD is caused by amyloid plaques and neurofibrillary tangles. In AD, β-amyloid plaques snipped from an amyloid precursor protein accumulate between neurons, which would be otherwise broken down and removed in a healthy brain. Again, neurofibrillary tangles primarily consist of the protein tau and surface as insoluble twisted fibers to be found inside the neurons of an AD brain. Tau is part of the microtubule structure, which enables the transportation of nutrients within a neuron. In AD, tau is abnormal and the microtubule structure is damaged (see Fig. 11.1). The systematic expansion lets us suggest that neuronal transport mechanisms spread these proteopathic seeds.

The neural substrates of MCI and AD have been extensively examined with electrophysiological and neuroimaging techniques. Words that study participants do not integrate semantically evoke a negative wave (N400) that peaks about 400 ms after the word's onset. Syntactic processes seemed to correspond to two different event-related potentials (ERP) components. The left anterior negativity (LAN) occurs in an early time window between 100 and 500 ms and the late centro-parietal positivity (P600) in the range of 500–1,000 ms. While LAN or early LAN corresponds to word category and morpho-syntactic errors, the P600 corresponds to processing complex sentence structures and resolving syntactic ambiguities (e.g., Kutas and Van Petten 1994). ERP studies as well as lexical priming studies reveal that sentence comprehension can be subdivided into different process stages. At an early process stage during online sentence comprehension, about 100 ms poststimulus onset, the listener automatically accesses lexical information; during the second stage, the listener integrates lexical information into a higher ordered sentence and discourse context. This process occurs in the range of about 400–1,000 ms poststimulus onset. ERP abnormalities in AD are prominent from latencies of approximately 200 ms and later. In contrast, sensory-dependent evoked potentials, such as N100, are generally normal in AD. Despite being applied to AD for about 25 years

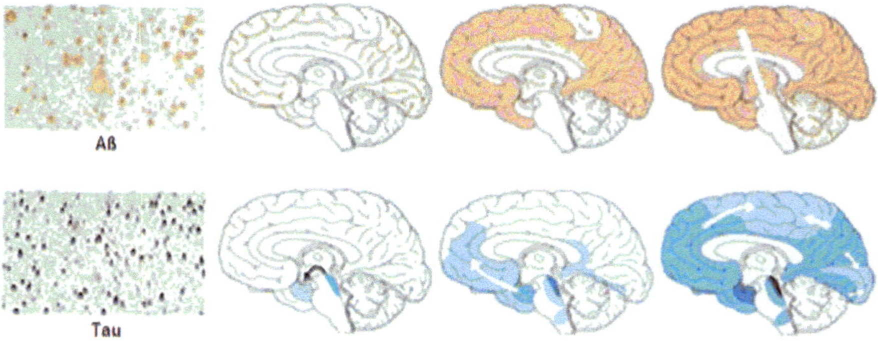

Fig. 11.3 The accumulation of proteins in AD is systematic. Cross-sectional autopsy indicates that β-amyloid plaques (*above*) first appear in the neocortex, followed by the allocortex (hippocampus and olfactory cortex), and finally in subcortical regions. Neurofibrillary tangles tau (*below*), occur first in the locus coeruleus and transentorhinal area, and then spread to the amygdala and interconnected neocortical brain regions. (Adapted and modified, Jucker and Walker 2011; © 2011 American Neurological Association)

since the early P300 studies, the full potential of ERPs in helping to diagnose MCI subjects and/or treat AD patients is yet to be realized (Golob et al. 2002, Fig. 11.3).

Let us briefly recall: Functional MRI studies, which used a word-level paradigm, indicate that the left MTG, the STG, the angular gyrus, and the left IFG are involved in lexical–semantic processing. Again, sentence processing studies indicate that the left IFG (BAs 45 & 47), the right STG, and the left MTG support semantic processing. Activations within BAs 45 and 47 were found when subjects performed an explicit judgment task that required the use of WM resources. It is therefore assumed that semantic processing occurs not only within the left temporal region, but also in the left frontal cortex, if explicit manipulation of semantic information is required. In the domain of syntax, functional neuroanatomical studies reveal that complex sentences are accompanied by increased activation within Broca's area (BAs 44 & 45). A differentiation of semantic and syntactic WM resources on anatomical basis seems to be premature. While the left IFG is obviously involved in explicit WM functions, temporal regions seem to be responsible for implicit semantic–syntactic processes. There are only few data on the neuroanatomic basis of language deficits in AD and MCI (e.g., Grossman et al. 2003). It is generally expected that in AD or MCI patients, no significant blood-oxygen level-dependent (BOLD) responses will be measured in the region of interest. However, in Saykin et al.'s (1999) fMRI study, AD patients showed in a phonological task additional foci in the left dorsolateral prefrontal and bilateral cingulate areas. In their semantic task, predominant activation foci were seen in the inferior and middle frontal gyrus (left greater than right), but AD patients showed additional activation suggesting compensatory recruitment of locally expanded foci and remote regions, for example in the right frontal region. AD is a progressive disease and longitudinal studies are particularly important for predicting the progress of the disease to provide effective interventions at different cognitive levels including language. In contrast to AD, which can affect most cognitive domains, autism spectrum disorders (ASDs)[3] are early developmental disorders, while other cognitive functions seem to be relatively spared.

ASD subjects are typically impaired in mentalizing, communicating, and interacting socially, but their specific neurological cause(s) are unknown.[4] Boys are five times more often affected by this disorder than girls—in contrast to Rett syndrome, which exclusively occurs in girls. In 2012, the incident rate of ASD in the USA was 1 in 88 children, but now it is rising. Different tasks were developed to reveal autistic symptoms in early childhood. The first-order tests imply our intuitive understanding that different people have different views of the same situation. Children with autism have difficulties to report what someone else thinks, but they simply tell what they themselves know (Swettenham et al. 1996). Similarly, by the age of 4 years, normal developing children typically can differentiate between mental words (e.g., *hope, know, think*) and nonmental words (e.g., *eat, jump, move*), but not autistic children. Moreover, young autistic children have difficulties to understand and produce deception, that is, to differentiate between someone's intention and the actual outcome, to interpret gaze direction, to understand emotional states and expressions, or to draw inferences between and seeing and knowing (e.g., Leslie ancd Frith 1988; Butterworth and Jarrett 1991; Baron-Cohen et al. 1993; Yirmiya et al. 1996; Castelli 2005).

Other first-order tasks, which are difficult for autistic children, include the use of imagination or comprehension of figurative and pragmatic language. For example, ASD children are less able to draw (usually) nonexisting, unreal objects such as a two-headed person (Scott and Baron-Cohen 1996). Comprehension of figurative language requires anticipating the speaker's intentions. Understanding a sarcastic statement such as *Your room is r e a l l y very clean!* typically starts by the age of 8 years. However, children with high-functioning autism may have difficulties understanding the speaker's intentions whether it is a sarcastic statement or a joke, and tend to take these figures of speech in a nonfigurative, literal sense. Also, autistic children seem unable to consider a conversation as a social-communicative process; for example, they have difficulties in recognizing a faux pas or adjusting their language behavior to the listener (Happé 1994; Baron-Cohen 1997).

In sum, the ability to interfere another person's mental state (first-order theory of mind: 4-year old mental state) is typically delayed in 4–5-year olds with high-functioning autism or Asperger syndrome. Because of this delay, these children often do not pass second-order tasks (Sperber and Wilson 1987). Tasks, which require recognizing indirect mental states such as *What X believes that Y believes* (second-order theory of mind (ToM): 6-year old mental state) may be passed by high-functioning

[3] Several neurological disorders are associated with symptoms of autism such as fragile X syndrome (changes of the gene FMR1), cerebral dysgenesis, Rett syndrome, or metabolic disorders.

[4] In California, high-risk regional clusters of autism were identified for the period from 1993 to 2001 (Mazumdar et al. 2010). These clusters seemed to be associated with the parents' higher educational level or with the proximity to a major treatment center for autism. Better education as well as local proximity to a center may raise awareness of this disorder. Environmental causes for these autism clusters could not be found. Also, scientific studies could not support the assumption that vaccination causes autism, an assumption widely held in the lay community.

autism/Asperger syndrome in their teens (Happé 1993, 1995). However, more subtle and complex tests may reveal mentalizing disorders even at older ages.

Structural MRI studies reveal atypical changes in the brain of ASD subjects. There is some evidence of an atypical brain growth in toddlers with autism. About 90% of autistic boys had a larger brain volume compared to controls (Courchesne et al. 2001): 3–4-year old toddler had significantly more white-matter volume but to a lesser extent more gray matter volume, but in the cerebellum, gray matter was reduced. Moreover, evidence has been reported that the brain growth halts in autism during childhood and it possibly degenerates during adulthood. In particular, in one study, 67% more neurons were found in the prefrontal cortex in children with autism (Courchesne et al. 2011). The prefrontal cortex supports mental functions such as planning, execution, and directing attention. This finding seems to match the behavioral deficits of autistic subjects in social cognition, but supports also the assumption of a genetic base of autism rather than being the result of postnatal events. However, there are other results, which question this interpretation. Macrocephaly seems not to reflect a homogenous group in ASD and applies only to a certain subgroup of autistic children (Lainhart et al. 1997; Fombonne et al. 1999). More recent neuroimaging studies of ASD confirm atypical local and global connectivity in fronto-temporal regions, which affect not only white-matter but also the intrinsic gray matter architecture (Ecker et al. 2013).

Another idea to explain autistic behavior refers to the assumption of a broken mirror neuron system (MNS). This is a plausible approach as the MNS encodes not only motor features but the goal of an action. Goal encoding is directly related to understand other people and to imitating them. Moreover, the MNS might thus also contribute to the ToM and to language (Iacoboni and Dapretto 2006; Gallese 2007, 2008; Oberman and Ramachandran 2007). Accordingly, the broken mirror theory of autism claims that poor social cognition in autism is related to a dysfunction of the MNS. Again, poor social cognition may have multidimensional effects on behavior, including pragmatic aspects of language. But which kind of evidence has been reported for the broken MNS in autism? Neuroimaging evidence reveals that typical children show in comparison to ASD children imitation tasks (with and without emotional aspects) stronger activation in the IFG and/or amygdala (e.g., Iacoboni 1999; Dapretto et al. 2006). However, other fMRI studies did not report differences between ASD and control subjects (e.g., Schulte-Rüther et al. 2011). The IFG is part of the core MNS and would thus support the broken mirror theory (Fig. 11.4). However, in reviewing the literature, the evidence can be regarded as inconclusive (see for a review, Hamilton 2013).

In sum, there is no clear evidence for the hypothesis of a broken mirror neuron system in autism. The atypical development of the prefrontal cortex, in which only BA 44 is considered as part of the MNS, speaks against the assumption that the dysfunction of mirror neurons plays an exclusive role in autism. The neurobiological causes of impaired social cognition in ASD seem to be related to the idea that the blueprint of the brain's development has been altered in autism. Furthermore, other cortical regions outside of prefrontal cortex (and of the mirror neuron system) are often also affected in autism. In ToM tasks, ASD subjects

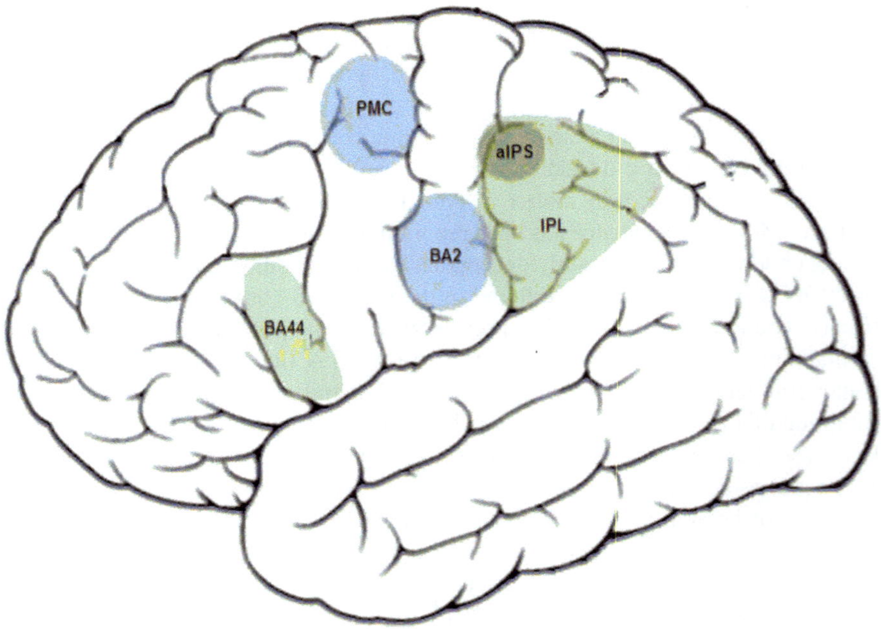

Fig. 11.4 Core regions of the human MNS include BA 44 (part of the IFG), the inferior parietal lobus (IPL) and the anterior intraparietal sulcus (aIPS). The primary motor cortex (PMC) and BA2 (an area of the somatosensory cortex) are both part of the extended mirror neuron system. (Adapted and modified, Hamilton 2013; © Elsevier Limited)

show degraded activation not only of the medial frontal cortex, the fusiform face area in the extrastriate cortex (BAs 18 & 19), but also of the superior temporal gyrus, atypical structures in the basal ganglia, cerebrellum, amgydala, hippocampus, corpus callosum and brain stem. From a theoretical viewpoint, it has been suggested that the autistic brain lacks top-down modulation of early sensory processing, which is caused by atypical connectivity and lack of pruning (Frith 2003). Although ASD is not a homogenous group, in general, it can be stated that atypical neural wiring affects language at the pragmatic level involving social and communicative abilities such as figurative speech. Core linguistic processes usually remain unaffected in ASD.

References

Alexander, M. P., Naeser, M. A., & Palumbo, C. (1990). Broca's area aphasias: Aphasia after lesions including the frontal operculum. *Neurology, 40,* 353–362.

Almor, A., Kempler, D., MacDonald, M. C., Andersen, E. S., & Tyler, L. K. (1999). Why do Alzheimer patients have difficulty with pronouns? Working memory, semantics, and reference in comprehension and production in Alzheimer's disease. *Brain and Language, 67*(3), 202–227.

Appell, J., Kertesz, A., & Fisman, M. (1982). A study of language functioning in Alzheimer patients. *Brain and Language, 17*(1), 73–91.

Balota, D. A., & Duchek, J. M. (1991). Semantic priming effects, lexical repetition effects, and contextual disambiguation effects in healthy aged individuals and individuals with senile dementia of the Alzheimer type. *Brain and Language, 40,* 181–201.

Baron-Cohen, S. (1997). Hey! It was just a joke! Understanding propositions and propositional attitudes by normally developing children and children with autism. *Israel Journal of Psychiatry and Related Sciences, 34*(3), 174–178.

Baron-Cohen, S., Spitz, A., & Cross, P. (1993). Can children with autism recognize surprise? *Cognition and Emotion, 7,* 507–516.

Bayles, K. A. (1982). Language function in senile dementia. *Brain and Language, 16*(2), 265–280.

Bayles, K. A., & Tomoeda, C. K. (1983). Confrontation naming impairment in dementia. *Brain and Language, 19*(1), 98–114.

Bayles, K. A., Tomoeda, C. K., & Trosset, M. W. (1990). Naming and categorical knowledge in Alzheimer's disease: The process of semantic memory deterioration. *Brain and Language, 39*(4), 498–510.

Baynes, K., Tramo, M. J., & Gazzaniga, M. S. (1992). Reading with a limited lexicon in the right hemisphere of a callosotomy patient. *Neuropsychologia, 30*(2), 187–200.

Bickel, C., Pantel, J., Eysenbach, K., & Schröder, J. (2000). Syntactic comprehension deficits in Alzheimer's disease. *Brain and Language, 71*(3), 432–448.

Binder, J. R., Frost, J. A., Hammeke, T. A., Cox, R. W., Rao, S. M., & Prieto, T. (1997). Human brain language areas identified by functional magnetic resonance imaging. *The Journal of Neuroscience, 17*(1), 353–362.

Blumstein, S. E., Milberg, W., & Shrier, R. (1982). Semantic processing in aphasia: Evidence from an auditory lexical decision task. *Brain and Language, 17*(2), 301–315.

Bookheimer, S. (2002). Functional MRI of language: New approaches to understanding the cortical organization of semantic processing. *Annual Review of Neuroscience, 25,* 151–188.

Bottini, G., Corcoran, R., Sterzi, R., Paulesu, E., Schenone, P., Scarpa, P., Frackowiak, R. S., & Frith, C. D. (1994). The role of the right hemisphere in the interpretation of figurative aspects of language. A positron emission tomography activation study. *Brain: A Journal of Neurology, 117*(6), 1241–1253.

Bradley, D. C. (1978). *Computational distinctions of vocabulary type. Unpublished doctoral dissertation.* Cambridge: MIT Press.

Broca, P. (1861). Remarques sur le siège de la faculté du langage articulé suivies d'une observation d'aphèmie (perte de la parole) [French]. *Bulletin de la Société d'Anatomie de Paris, 6,* 330–357.

Butters, N., Granholm, E., Salmon, D. P., Grant, I., & Wolfe, J. (1987). Episodic and semantic memory: A comparison of amnesic and demented patients. *Journal of Clinical and Experimental Neuropsychology, 9*(5), 479–497.

Butterworth, G., & Jarrett, N. (1991). What minds have in common is space: Spatial mechanisms serving joint visual attention in infancy. *British Journal of Developmental Psychology, 9*(1), 55–72.

Caplan, D. (1987). *Neurolinguistics and linguistic aphasiology: An introduction.* Cambridge: Cambridge University Press.

Caplan, D., & Hildebrandt, N. (1988). *Disorders of syntactic comprehension.* Cambridge: MIT Press.

Caramazza, A., & Zurif, E. B. (1976). Dissociation of algorithmic and heuristic processes in language comprehension: Evidence from aphasia. *Brain and Language, 3*(4), 572–582.

Caramazza, A., Berndt, R. S., & Brownell, H. H. (1982). The semantic deficit hypothesis: Perceptual parsing and object classification by aphasic patients. *Brain and Language, 15*(1), 161–189.

Castelli, F. (2005). Understanding emotions from standardized facial expressions in autism and normal development. *Autism: The International Journal of Research and Practice, 9*(4), 428–449.

Chertkow, H., & Bub, D. (1990). Semantic memory loss in dementia of Alzheimer's type. What do various measures measure? *Brain: A Journal of Neurology, 113* (2), 397–417.

Chertkow, H., Bub, D., & Seidenberg, M. (1989). Priming and semantic memory loss in Alzheimer's disease. *Brain and Language, 36,* 420–446.

Corder, E. H., Saunders, A. M., Risch, N. J., Strittmatter, W. J., Schmechel, D. E., Gaskell, P. C., Jr, J., Rimmler, B., Locke, P. A., Conneally, P. M., Schmader, K. E., Small, G. W., Roses, A. D., Haines, J. L., & Pericak-Vance, M. A. (1994). Protective effect of apolipoprotein E type 2 allele for late onset Alzheimer disease. *Nature Genetics, 7*(2), 180–184.

Cornelissen, K., Laine, M., Tarkiainen, A., Järvensivu, T., Martin, N., & Salmelin, R. (2003). Adult brain plasticity elicited by anomia treatment. *Journal of Cognitive Neuroscience, 15*(3), 444–461.

Courchesne, E., Karns, C. M., Davis, H. R., Ziccar di, R., Carper, R. A., Tigue, Z. D., Chisum, H. J., Moses, P., Pierce, K., Lord, C., Lincoln, A. J., Pizzo, S., Schreibman, L., Haas, R. H., Akshoomoff, N. A., & Courchesne, R. Y. (2001). Unusual brain growth patterns in early life in patients with autistic disorder: An MRI study. *Neurology, 57*(2), 245–254.

Courchesne, E., Mouton, P. R., Calhoun, M. E., Semendeferi, K., Ahrens-Barbeau, C., Hallet, M. J., Barnes, C. C., & Pierce, K. (2011). Neuron number and size in prefrontal cortex of children with autism. *JAMA: The Journal of the American Medical Association, 306*(18), 2001–2010.

Cross, K., Smith, E. E., & Grossman, M. (2008). Knowledge of natural kinds in semantic dementia and Alzheimer's disease. *Brain and Language, 105*(1), 32–40.

Cummings, J. L., Benson, F., Hill, M. A., & Read, S. (1985). Aphasia in dementia of the Alzheimer type. *Neurology, 35*(3), 394–397.

Damasio, H., Grabowski, T. J., Tranel, D., Hichwa, R. D., & Damasio, A. R. (1996). A neural basis for lexical retrieval. *Nature, 380*(6574), 499–505.

Dapretto, M., Davies, M. S., Pfeifer, J. H., Scott, A. A., Sigman, M., Bookheimer, S. Y., & Iacoboni, M. (2006). Understanding emotions in others: Mirror neuron dysfunction in children with autism spectrum disorders. *Nature Neuroscience, 9*(1), 28–30.

De Renzi, E., & Vignolo, L. A. (1962). The token test: A sensitive test to detect receptive disturbances in aphasics. *Brain: A Journal of Neurology, 85*(4), 665–678.

Dronkers, N. F. (1996). A new brain region for coordinating speech articulation. *Nature, 384,* 159–161.

Ecker, C., Ronan, L., Feng, Y., Daly, E., Murphy, C., Ginestet, C. E., Brammer, M., Fletcher, P. C., Bullmore, E. T., Suckling, J., Baron-Cohen, S., Williams, S., Loth, E., MRC AIMS Consortium, & Murphy, D. G. (2013). Intrinsic gray-matter connectivity of the brain in adults with autism spectrum disorder. *Proceedings of the National Academy of Sciences, 110*(32), 13222–13227.

Emery, O. B., & Breslau, L. D. (1989). Language deficits in depression: Comparisons with SDAT and normal aging. *Journal of Gerontology, 44*(3), M85–92.

Fodor, J. A. (1983). *The modularity of mind.* Cambridge: MIT Press.

Fombonne, E., Rogé, B., Claverie, J., Courty, S., & Frémolle, J. (1999). Microcephaly and macrocephaly in autism. *Journal of Autism and Developmental Disorders, 29*(2), 113–119.

Forster, K. I. (1979). Levels of processing and the structure of the language processor. In W. E. Cooper & E. Walker (Eds.), *Sentence processing: Psycholinguistics studies presented to Merrill Garret.* Hillsdale: Larence Erlbaum.

Frackowiak, R. S. (2001). New functional cerebral cartography: Studies of plasticity of the human brain. *Bulletin de l'Académie nationale de médecine, 185*(4), 707–724.

Frith, C. (2003). What do imaging studies tell us about the neural basis of autism? *Novartis Foundation Symposium, 251,* 149–166.

Gainotti, G., Caltagirone, C., Miceli, G., & Masullo, C. (1981). Selective semantic-lexical impairment of language comprehension in right-brain-damaged patients. *Brain and Language, 13*(2), 201–211.

Gallese, V. (2007). Before and below "theory of mind": Embodied simulation and the neural correlates of social cognition. *Philosophical Transactions of the Royal Society B: Biological Sciences, 362*(1480), 659–669.

Gallese, V. (2008). Mirror neurons and the social nature of language: The neural exploitation hypothesis. *Social Neuroscience, 3*(3–4), 317–333.

Gazzaniga, M. S. (1995). Principles of human brain organization derived from split-brain studies. *Neuron, 14*(2), 217–228.

Gazzaniga, M. S., & Sperry, R. W. (1967). Language after section of the cerebral commissures. *Brain: A Journal of Neurology, 90*(1), 131–148.

Golob, E. J., Johnson, J. K., & Starr, A. (2002). Auditory event-related potentials during target detection are abnormal in mild cognitive impairment. *Clinical Neurophysiology, 113*(1), 151–161.

Goodglass, H. (1976). Agrammatism. In H. Whitaker & H. A. Whitaker (Eds.), *Studies in neurolinguistics, Vol. 1*. New York: Academic Press.

Goodglass, H., & Baker, E. (1976). Semantic field, naming, and auditory comprehension in aphasia. *Brain and Language, 3,* 359–374.

Goodglass, H., & Berko, J. (1960). Agrammatism and inflectional morphology in English. *Journal of Speech and Hearing Research, 3,* 257–267.

Gordon, B., & Caramazza, A. (1983). Closed- and open-class lexical access in agrammatic and fluent aphasics. *Brain and Language, 19*(2), 335–345.

Grober, E., Buschke, H., Kawas, C., & Fuld, P. (1985). Impaired ranking of semantic attributes in dementia. *Brain and Language, 26*(2), 276–286.

Grossman, M., & Rhee, J. (2001). Cognitive resources during sentence processing in Alzheimer's disease. *Neuropsychologia, 39*(13), 1419–1431.

Grossman, M., & White-Devine, T. (1998). Sentence comprehension in Alzheimer's disease. *Brain and Language, 62*(2), 186–201.

Grossman, M., Mickanin, J., Onishi, K., & Hughes, E. (1995). An aspect of sentence processing in Alzheimer's disease: Quantifier-noun agreement. *Neurology, 45*(1), 85–91.

Grossman, M., Mickanin, J., Robinson, K. M., & D'Esposito, M. (1996). Anomaly judgments of subject-predicate relations in Alzheimer's disease. *Brain and Language, 54*(2), 216–232.

Grossman, M., Robinson, K., Biassou, N., White-Devine, T., & D'Esposito, M. (1998). Semantic memory in Alzheimer's disease: Representativeness, ontologic category, and material. *Neuropsychology, 12*(1), 34–42.

Grossman, M., Koenig, P., DeVita, C., Glosser, G., Moore, P., Gee, J., Detre, J., & Alsop, D. (2003). Neural basis for verb processing in Alzheimer's disease: An fMRI study. *Neuropsychology, 17*(4), 658–674.

Hamilton, A. F. de C. (2013). Reflecting on the mirror neuron system in autism: A systematic review of current theories. *Developmental Cognitive Neuroscience, 3,* 91–105.

Happé, F. G. (1993). Communicative competence and theory of mind in autism: A test of relevance theory. *Cognition, 48*(2), 101–119.

Happé, F. G. (1994). An advanced test of theory of mind: Understanding of story characters' thoughts and feelings by able autistic, mentally handicapped, and normal children and adults. *Journal of Autism and Developmental Disorders, 24*(2), 129–154.

Happé, F. G. (1995). The role of age and verbal ability in the theory of mind task performance of subjects with autism. *Child Development, 66*(3), 843–855.

Heiss, W. D., Kessler, J., Thiel, A., Ghaemi, M., & Karbe, H. (1999). Differential capacity of left and right hemispheric areas for compensation of poststroke aphasia. *Annals of Neurology, 45*(4), 430–438.

Hier, D. B., Hagenlocker, K., & Shindler, A. G. (1985). Language disintegration in dementia: Effects of etiology and severity. *Brain and Language, 25*(1), 117–133.

Hillert, D. (1990). *Sprachprozesse und Wissensstrukturen* [German]. Opladen: Westdeutscher Press.

Hillert, D. (1999). On processing lexical concepts in aphasia and Alzheimer's disease. Some (Re) considerations. *Brain and Language, 69,* 95–118.

Hillert, D. G. (2004). Spared access to idiomatic and literal meanings: A single-case approach. *Brain and Language, 89*(1), 207–215.

Hillis, A. E. (2002). Does the right make it right? Questions about recovery of language after stroke. *Annals of Neurology, 51*(5), 537–538.

Hodges, J. R., Salmon, D. P., & Butters, N. (1992). Semantic memory impairment in Alzheimer's disease: Failure of access or degraded knowledge? *Neuropsychologia, 30*(4), 301–314.

Huff, F. J., Corkin, S., & Growden, J. H. (1986). Semantic impairment and anomia in Alzheimer's disease. *Brain and Language, 28,* 235–249.

Iacoboni, M. (1999). Cortical mechanisms of human imitation. *Science, 286*(5449), 2526–2528.

Iacoboni, M., & Dapretto, M. (2006). The mirror neuron system and the consequences of its dysfunction. *Nature Reviews Neuroscience, 12,* 942–951.

Ingram, J. C. L. (2007). *Neurolinguistics: An introduction to spoken language processing and its disorders.* Cambridge: Cambridge University Press.

Jucker, M., & Walker, L. C. (2011). Pathogenic protein seeding in Alzheimer disease and other neurodegenerative disorders. *Annals of Neurology, 70*(4), 532–540.

Katz, W. F. (1988). An investigation of lexical ambiguity in Broca's aphasics using an auditory lexical priming technique. *Neuropsychologia, 26*(5), 747–752.

Kean, M.-L. (1977). The Linguistic Description of Aphasic Syndromes. *Cognition, 5,* 9–46.

Kemper, S., LaBarge, E., Ferraro, F. R., Cheung, H., Cheung, H., & Storandt, M. (1993). On the preservation of syntax in Alzheimer's disease. Evidence from written sentences. *Archives of Neurology, 50*(1), 81–86.

Kempler, D., Almor, A., Tyler, L. K., Andersen, E. S., & MacDonald, M. C. (1998). Sentence comprehension deficits in Alzheimer's disease: A comparison of off-line vs. on-line sentence processing. *Brain and Language, 64*(3), 297–316.

Kertesz, A., Appell, J., & Fisman, M. (1986). The dissolution of language in Alzheimer's disease. *The Canadian Journal of Neurological Sciences. Le Journal Canadien Des Sciences Neurologiques, 13*(4), 415–418.

Kertesz, A., Lau, W. K., & Polk, M. (1993). The structural determinants of recovery in Wernicke's aphasia. *Brain and Language, 44,* 153–164.

Kolk, H. H., & Blomert, L. (1985). On the Bradley hypothesis concerning agrammatism: The nonword-interference effect. *Brain and Language, 26*(1), 94–105.

Kussmaul, A. (1881). *Die Störungen der Sprache.* Leipzig: F. C. W. Vogel [In German].

Kutas, M., & Van Petten, C. (1994). Psycholingusistics electrified: Event-related brain potential investigations. In M. Gernsbacher (Ed.), *Handbook of psycholinguistics* (pp. 83–143). New York: Academic Press.

Lainhart, J. E., Piven, J., Wzorek, M., Landa, R., Santangelo, S. L., Coon, H., & Folstein, S. E. (1997). Macrocephaly in children and adults with autism. *Journal of the American Academy of Child and Adolescent Psychiatry, 36*(2), 282–290.

Leslie, A. M., & Frith, U. (1988). Autistic children's understanding of seeing, knowing and believing. *British Journal of Developmental Psychology, 6*(4), 315–324.

Lesser, R. (1978). *Linguistic investigations of aphasia.* New York: Elsevier.

Lichtheim, L. (1884). Ueber Aphasie [German]. *Deutsches Archiv Für Klinische Medicin, 36,* 204–268.

Lopez, O. L., Jagust, W. J., DeKosky, S. T., Becker, J. T., Fitzpatrick, A., Dulberg, C., Breitner, J., Lyketsos, C., Jones, B., Kawas, C., Carlson, M., & Kuller, L. H. (2003). Prevalence and classification of mild cognitive impairment in the cardiovascular health study cognition Study, Part 1. *Archives of Neurology, 60,* 1385–1389.

Lukatela, K., Crain, S., & Shankweiler, D. (1988). Sensitivity to inflectional morphology in agrammatism: Investigation of a highly inflected language. *Brain and Language, 33,* 1–15.

Martin, R. C. (1987). Articulatory and phonological deficits in short-term memory and their relation to syntactic processing. *Brain and Language, 32*(1), 159–192.

Martin, A., & Fedio, P. (1983). Word production and comprehension in Alzheimer's disease: The breakdown of semantic knowledge. *Brain and Language, 19*(1), 124–141.

Mazumdar, S., King, M., Liu, K.-Y., Zerubavel, N., & Bearman, P. (2010). The spatial structure of autism in California, 1993-2001. *Health & Place, 16*(3), 539–546.

McClelland, J. L. (1987). The case for interactionism in language processing. In M. Coltheart (Ed.), *Attention & performance XII: The psychology of reading* (pp. 1–36). London: Erlbaum.

Meinzer, M., Flaisch, T., Breitenstein, C., Wienbruch, C., Elbert, T., & Rockstroh, B. (2008). Functional re-recruitment of dysfunctional brain areas predicts language recovery in chronic aphasia. *NeuroImage, 39*(4), 2038–2046.

Mesulam, M. (1994). Neurocognitive networks and selectively distributed processing. *Revue Neurologique, 150*(8-9), 564–569.

Miceli, G., Mazzucchi, A., Menn, L., & Goodglass, H. (1983). Contrasting cases of Italian agrammatic aphasia without comprehension disorder. *Brain and Language, 19*(1), 65–97.

Milberg, W., & Blumstein, S. E. (1981). Lexical decision and aphasia: Evidence for semantic processing. *Brain and Language, 14*(2), 371–385.

Milberg, W., Blumstein, S. E., & Dworetzky, B. (1987). Processing of lexical ambiguities in aphasia. *Brain and Language, 31*(1), 138–150.

Mohr, J. P. (1976). Broca's area and Broca's aphasia. In H. Whitaker & H. A. Whitaker (Eds.), *Studies in neurolinguistics, Vol. 1*. New York: Academic Press.

Molloy, R., Brownell, H. H., & Gardner H. (1990). Discourse comprehension by right-hemisphere stroke patients: Deficits in prediction and revision. In Y. Joanette & H. M. Brownell (Eds.), *Discourse ability and brain damage: Theoretical and empirical perspectives* (pp. 113–130). New York: Springer Press.

Morris, J. C., Storandt, M., Miller, J. P., McKeel, D. W., Price, J. L., Rubin, E. H., & Berg, L. (2001). Mild cognitive impairment represents early-stage Alzheimer disease. *Archive of Neurology, 58*(3), 397–405.

Murdoch, B. E., Afford, R. J., Ling, A. R., & Ganguley, B. (1986). Acute computerized tomographic scans: Their value in the localization of lesions and as prognostic indicators in aphasia. *Journal of Communication Disorders, 19*(5), 311–345.

Murdoch, B. E., Chenery, H. J., Wilks, V., & Boyle, R. S. (1987). Language disorders in dementia of the Alzheimer type. *Brain and Language, 31*(1), 122–137.

Musso, M., Weiller, C., Kiebel, S., Müller, S. P., Bülau, P., & Rijntjes, M. (1999). Training-induced brain plasticity in aphasia. *Brain: A Journal of Neurology, 122*(9), 1781–1790.

Myers, P. S., & Linebaugh, C. W. (1981). Comprehension of idiomatic expressions by right hemisphere damaged adults. In R. H. Brookshire (Ed.), *Clinical Aphasiology* (pp. 254–261). Minneapolis: BRK Publishers.

Naeser, M. A., & Palumbo, C. L. (1994). Neuroimaging and language recovery in stroke. *Journal of Clinical Neurophysiology: Official Publication of the American Electroencephalographic Society, 11*(2), 150–174.

Nebes, R. D., Boller, F., & Holland, A. (1986). Use of semantic context by patients with Alzheimer's disease. *Psychology and Aging, 1*, 261–269.

Nebes, R. D., Brady, C. B., & Huff, F. J. (1989). Automatic and attentional mechanisms of semantic priming in Alzheimer's disease. *Journal of Clinical and Experimental Neuropsychology, 11*(2), 219–230.

Nespoulous, J. L., Dordain, M., Perron, C., Ska, B., Bub, D., Caplan, D., Mehler, J., & Lecours, A. R. (1988). Agrammatism in sentence production without comprehension deficits: Reduced availability of syntactic structures and/or of grammatical morphemes? A case study. *Brain and Language, 33*(2), 273–295.

Nichelli, P., Grafman, J., Pietrini, P., Clark, K., Lee, K. Y., & Miletich, R. (1995). Where the brain appreciates the moral of a story. *Neuroreport, 6*(17), 2309–2313.

Nocentini, U., Goulet, P., Drolet, M., & Joanette, Y. (1999). Age-related evolution of the contribution of the right hemisphere to language: Absence of evidence. *The International Journal of Neuroscience, 99*(1-4), 59–67.

Ober, B. A., & Shenaut, G. K. (1988). Lexical decision and priming in Alzheimer's disease. *Neuropsychologia, 26*, 273–286.

Oberman, L. M., & Ramachandran, V. S. (2007). The simulating social mind: The role of the mirror neuron system and simulation in the social and communicative deficits of autism spectrum disorders. *Psychological Bulletin, 133*(2), 310–327.

Ohyama, M., Senda, M., Kitamura, S., Ishii, K., Mishina, M., & Terashi, A. (1996). Role of the nondominant hemisphere and undamaged area during word repetition in poststroke aphasics. A PET activation study. *Stroke: A Journal of Cerebral Circulation, 27*(5), 897–903.

Ostrin, R. K., & Tyler, L. K. (1993). Automatic access to lexical semantics in aphasia: Evidence from semantic and associative priming. *Brain and Language, 45*(2), 147–159.

Peng, F. C. C. (2009). *Language in the brain: Critical assessments*. London: Continuum Intl Pub Group.
Petersen, R. C. (2003). Conceptual overview. In R. C. Petersen (Ed.), *Mild cognitive impairment: Aging to Alzheimer's disease*. New York: Oxford University Press.
Pick, A. (1913). *Die agrammatischen Sprachstörungen*. Berlin: Springer.
Poldrack, R. A., Wagner, A. D., Prull, M. W., Desmond, J. E., Glover, G. H., & Gabrieli, J. D. (1999). Functional specialization for semantic and phonological processing in the left inferior prefrontal cortex. *NeuroImage, 10*(1), 15–35.
Posner, M. I., & Snyder, C. R. R. (1975). Attention and cognitive control. In R. L. Solso (Ed.), *Information processing and cognition: The loyola symposium*. Hillsdale: Lawrence Erlbaum.
Postman-Caucheteux, W. A., Birn, R. M., Pursley, R. H., Butman, J. A., Solomon, J. M., Picchioni, D., McArdle, J., & Braun, A. R. (2010). Single-trial fMRI shows contralesional activity linked to overt naming errors in chronic aphasic patients. *Journal of Cognitive Neuroscience, 22*(6), 1299–1318.
Prather, P. A., Zurif, E., Love, T., & Brownell, H. (1997). Speed of lexical activation in nonfluent Broca's aphasia and fluent Wernicke's aphasia. *Brain and Language, 59*(3), 391–411.
Price, C. I., Green, D. W., & von Studnitz, R. (1999). A functional imaging study of translation and language switching. *Brain, 122,* 2221–2235.
Rochon, E., Waters, G. S., & Caplan, D. (1994). Sentence comprehension in patients with Alzheimer's disease. *Brain and Language, 46*(2), 329–349.
Rosen, H. J., Petersen, S. E., Linenweber, M. R., Snyder, A. Z., White, D. A., Chapman, L., Dromerick, A. W., Fiez, J. A., & Corbetta, M. D. (2000). Neural correlates of recovery from aphasia after damage to left inferior frontal cortex. *Neurology, 55*(12), 1883–1894.
Roskies, A. L., Fiez, J. A., Balota, D. A., Raichle, M. E., & Petersen, S. E. (2001). Task-dependent modulation of regions in the left inferior frontal cortex during semantic processing. *Journal of Cognitive Neuroscience, 13*(6), 829–843.
Sakai, K. L., Miura, K., Narafu, N., & Muraishi, Y. (2004). Correlated functional changes of the prefrontal cortex in twins induced by classroom education of second language. *Cereb Cortex, 14,* 1233–1239.
Salmon, D. P., Shimamura, A., Butters, N., & Smith, S. (1988). Lexical and semantic priming deficits in patients with Alzheimer's disease. *Journal of Clinical and Experimental Neuropsychology, 10,* 477–494.
Saur, D., Lange, R., Baumgaertner, A., Schraknepper, V., Willmes, K., Rijntjes, M., & Weiller, C. (2006). Dynamics of language reorganization after stroke. *Brain: A Journal of Neurology, 129*(6), 1371–1384.
Saykin, A. J., Flashman, L. A., Frutiger, S. A., Johnson, S. C., Mamourian, A. C., Moritz, C. H., O'Jile, J. R., Riordan, H. J., Santulli, R. B., Smith, C. A., & Weaver, J. B. (1999). Neuroanatomic substrates of semantic memory impairment in Alzheimer's disease: Patterns of functional MRI activation. *Journal of the International Neuropsychological Society: JINS, 5*(5), 377–392.
Schlaug, G., Marchina, S., & Norton, A. (2009). Evidence for plasticity in white-matter tracts of patients with chronic Broca's aphasia undergoing intense intonation-based speech therapy. *Annals of the New York Academy of Sciences, 1169,* 385–394.
Schneider, W., & Shiffrin, R. M. (1977). Controlled and automatic human information processing: 1. Detection, search, and attention. *Psychological Review, 84,* 1–66.
Schulte-Rüther, M., Greimel, E., Markowitsch, H. J., Kamp-Becker, I., Remschmidt, H., Fink, G. R., & Piefke, M. (2011). Dysfunctions in brain networks supporting empathy: An fMRI study in adults with autism spectrum disorders. *Social Neuroscience, 6*(1), 1–21.
Schwartz, M., Marin, O., & Saffran, E. (1979). Dissociations of language function in dementia: A case study. *Brain and Language, 7,* 277–306.
Schwartz, M. F., Linebarger, M. C., Saffran, E. M., & Pate, D. S. (1987). Syntactic transparency and sentence interpretation in aphasia. *Language and Cognitive Processes, 2*(2), 85–113.
Scott, F. J., & Baron-Cohen, S. (1996). Imagining real and unreal things: Evidence of a dissociation in autism. *Journal of Cognitive Neuroscience, 8*(4), 371–382.
Seidenberg, M. S. (1985). Lexicon as module. *The Behavioral and Brain Sciences, 8*(1), 31–32.

Selnes, O. A., Knopman, D. S., Niccum, N., Rubens, A. B., & Larson, D. (1983). Computed tomographic scan correlates of auditory comprehension deficits in aphasia: A prospective recovery study. *Annals of Neurology, 13*(5), 558–566.

Shiffrin, R. M., & Schneider, W. (1977). Controlled and automatic human information processing: II. Perceptual learning, automatic attending, and a general theory. *Psychological Review, 84*(127), 190.

Sperber, D., & Wilson, D. (1987). Precis of relevance: Communication and cognition. *Behavioral and Brain Sciences., 10,* 697–754.

Sperber, D. (2001). *Relevance: Communication and cognition* (2nd edn.). Oxford: Blackwell Publishers.

Suh, S., Yoon, H. W., Lee, S., Chung, J.-Y., Cho, Z.-H., & Park, H. (2007). Effects of syntactic complexity in L1 and L2; An fMRI study of Korean–English bilinguals. *Brain Research, 1136,* 178–189.

Swettenham, J. G., Baron-Cohen, S., Gomez, J. C., & Walsh, S. (1996). What's inside someone's head? Conceiving of the mind as a camera helps children with autism acquire an alternative to a theory of mind. *Cognitive Neuropsychiatry, 1*(1), 73–88.

Swinney, D., Zurif, E., & Nicol, J. (1989). The effects of focal brain damage on sentence processing: An examination of the neurological organization of a mental module. *Journal of Cognitive Neuroscience, 1*(1), 25–37.

Thompson, C. K. (2000). Neuroplasticity: Evidence from aphasia. *Journal of Communication Disorders, 33*(4), 357–366.

Thompson-Schill, S. L. (2003). Neuroimaging studies of semantic memory: Inferring "how" from "where.". *Neuropsychologia, 41*(3), 280–292.

Thompson-Schill, S. L., Aguirre, G. K., D'Esposito, M., & Farah, M. J. (1999). A neural basis for category and modality specificity of semantic knowledge. *Neuropsychologia, 37*(6), 671–676.

Thulborn, K. R., Carpenter, P. A., & Just, M. A. (1999). Plasticity of language-related brain function during recovery from stroke. *Stroke: A Journal of Cerebral Circulation, 30*(4), 749–754.

Tompkins, C. A., Boada, R., & McGarry, K. (1992). The access and processing of familiar idioms by brain-damaged and normally aging adults. *Journal of Speech and Hearing Research, 35*(3), 626–637.

Van Lancker, D. (1990). The neurology of proverbs. *Behavioural Neurology, 3,* 169–187.

Van Lancker, D. R., & Kempler, D. (1987). Comprehension of familiar phrases by left- but not by right-hemisphere damaged patients. *Brain and Language, 32*(2), 265–277.

Vandenberghe, R., Price, C., Wise, R., Josephs, O., & Frackowiak, R. S. (1996). Functional anatomy of a common semantic system for words and pictures. *Nature, 383*(6597), 254–256.

Wagner, A. D., Paré-Blagoev, E. J., Clark, J., & Poldrack, R. A. (2001). Recovering meaning: Left prefrontal cortex guides controlled semantic retrieval. *Neuron, 31*(2), 329–338.

Warburton, L. (1999). Management of stroke: A practical guide for the prevention, evaluation and treatment of acute stroke. *Journal of Neurology, Neurosurgery, and Psychiatry, 66*(5), 696A.

Waters, G., & Caplan, D. (2002). Working memory and online syntactic processing in Alzheimer's disease: Studies with auditory moving window presentation. *Journal of Gerontoly. Series B. Psychology Sciences Social Sciences, 57*(4), 298–311.

Waters, G. S., Rochon, E., & Caplan, D. (1998). Task demands and sentence comprehension in patients with dementia of the Alzheimer's type. *Brain and Language, 62*(3), 361–397.

Weber-Fox, C. M., & Neville, H. J. (1996). Maturational Constraints on Functional Specializations for Language Processing: ERP and Behavioral Evidence in Bilingual Speakers. *Journal of Cognitive Neuroscience, 8*(3), 231–256.

Weiller, C., Isensee, C., Rijntjes, M., Huber, W., Müller, S., Bier, D., Dutschka, K., Woods, R. P., Noth, J., & Diener, H. C. (1995). Recovery from Wernicke's aphasia: A positron emission tomographic study. *Annals of Neurology, 37*(6), 723–732.

Wernicke, C. (1874). *Der aphasische Symptomencomplex, eine psychologische Studie auf anatomischer Basis* [German]. Breslau: M. Cohn und Weigert.

Winner, E., & Gardner, H. (1977). The comprehension of metaphor in brain-damaged patients. *Brain: A Journal of Neurology, 100*(4), 717–729.

Yirmiya, N., Solomonica-Levi, D., & Shulman, C. (1996). The ability to manipulate behaviour and to understand manupulation of beliefs: A comparison of individuals with autism, mental retardation, and normal development. *Developmental Psychology, 32,* 62–69.

Zurif, E. B., & Caramazza, A. (1976). Dissociation of algorithmic and heuristic processes in language comprehension: Evidence from aphasia. *Brain and Language, 3*(4), 572–582.

Zurif, E., Swinney, D., Prather, P., Solomon, J., & Bushell, C. (1993). An on-line analysis of syntactic processing in Broca's and Wernicke's aphasia. *Brain and Language, 45*(3), 448–464.

Chapter 12
Language Acquisition

12.1 The Genetic Program

Different studies have demonstrated that newborns prefer to listen to the voice of their mothers rather than to the voice of other people (DeCasper and Fifer 1980) and they prefer their native language as compared to a different language (Moon et al. 1993). However, paying attention to voices seems to take place before birth as several studies indicate these effects also in fetuses (Kisilevsky et al. 2003, 2009). Thereby, the preference for the mother's speech does not reflect a preference for higher frequencies, as the mother voice preference effect is the same as compared to the father's voice or to a stranger's female or male voice (DeCasper and Prescott 1984). Thus, neural circuits sensitive to the mother's speech are shaped perinatally. It would be interesting to know whether the fetus would be sensitive to more than one language if the mother converses in more than one language.

During the perinatal period, the accumulation of synapses in the sensory, motor, and association cortices occurs most rapidly. According to an epigenetic model of language acquisition, genetic expressions of neural growth interact with species-specific experiences (Werker and Tees 1984). Similarly, it has been argued that the neural growth in the perinatal period involves the transition from gene expression to a experience–expectant phase, in which external ubiquitous stimuli common to all members of a species (e.g., speech recognition) are required to modify the neural circuits. Neurons, which are in excess, are pruned and dendritic attrition shapes the neural circuits. These circuits are thought to be relatively permanent, but can be used for subsequent neural sculpturing. During this phase, parameters are set for the critical or sensitive period phenomena. In contrast, during the experience–dependent phase, idiosyncratic information, which is unique to the individual, is stored. Thereby, active formation of new synaptic connections in response to the input will be stored (Greenough et al. 1987). The authors, however, also emphasize that it might be difficult dividing clearly both types of brain mechanisms involved in cognitive development. In sum, both mechanisms apparently work against each other: stable and optimized neural circuits are the result of neural pruning and at the same time new neural circuits can be developed to assimilate new information. Thus, the formation of particular neural circuits for speech processing seems to

D. Hillert, *The Nature of Language*, DOI 10.1007/978-1-4939-0609-3_12,

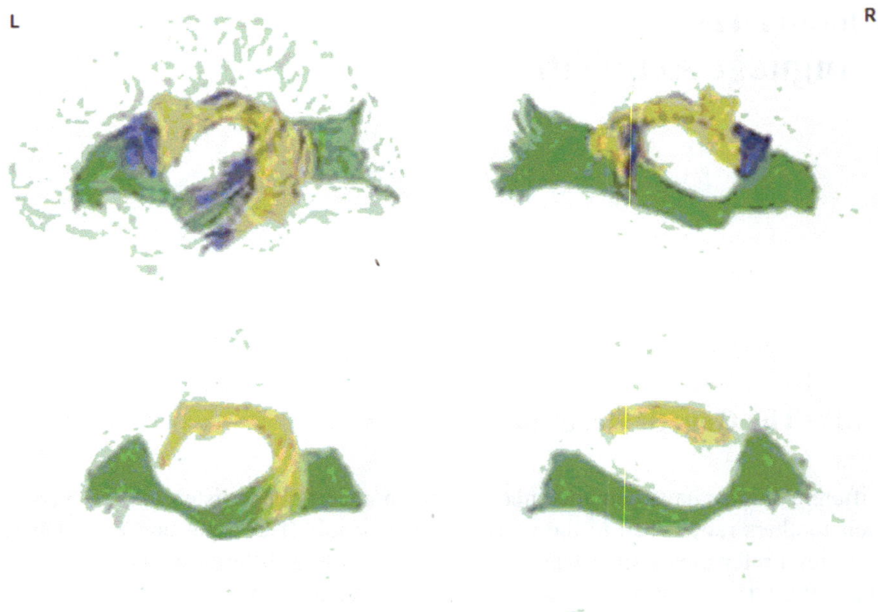

Fig. 12.1 Fiber tracking in adults (above) and newborns (below) by using MR-DTI. Two dorsal and one ventral pathway can be seen in adults and in newborns one dorsal and one ventral pathway. ■: The temporal lobe (TL) is dorsally connected via the fasciculus arcuates (AF) and the superior longitudinal fasciculus (SLF) to the inferior frontal gyrus (IFG) including Broca's area. ■: TL is dorsally connected via the AF & SLF to the premotor cortex. ■: The IFG is ventrally connected via the extreme capsule fiber to the TL (left/right hemisphere, *L*/*R*). (Adapted and modified, Perani et al. 2011; © Proceedings of the National Academy of Sciences)

follow a species-specific genetic program, while other circuits are shaped by neural plasticity to cope, for example, with new lexical information. The acquisition of syntax, however, might not rely entirely on one of both mechanisms. For example, a canonical structure, which can be directly mapped onto our sensory-motor experience, might be acquired by means of experience-expectant processes, while other syntactic structures primarily by experience-dependent processes. Whether both types of neural acquisition mechanisms are also reflected in the development of fiber tracks remains to be seen (see Fig. 12.1).

The child is not a passive learner as the readiness of language acquisition requires practice reflected in different periods of acquisition (after birth to the 3+ years): various sounds, repetitive CV patterns (babbling, 6–8 months), content words (holophrastic, 9–18 months.), two-word phrases with semantic relations (18–24 months), early multi-words (telegraphic, 24–30 months), late multi-word-phrases with morphemes (30+ months). Figurative language will be acquired relatively late. For instance, idiomatic expressions may be acquired between 6–10 years of age, if we ignore a few highly frequent expressions; irony and other more sophisticated figurative speech acts are acquired even later. One of the most discussed topics of language acquisition is the concept of a critical period.

The concept of a critical period (in contrast to a sensitive period) refers to a phase in the life span of an organism, in which it develops or acquires a particular skill. If the organism is not exposed to the relevant stimuli during this critical phase, it is difficult or even impossible to use these skills later in life. For example, the common chaffinch must be exposed to the songs of an adult chaffinch before adulthood, before it is sexually mature, to be able to acquire this intricate song. A critical period for language acquisition has been claimed by Lenneberg (1967) (see also Pinker 1994). Lenneberg argued that the critical language period is between 5 years of age and puberty, and referred to the observation that feral (e.g., “Genie”; see Rymer 1994) or deaf children have difficulties acquiring spoken language after puberty. Moreover, he assumed that children with neurologically caused language disorders recover significantly better and faster than adults with comparable impairments. This argument is, however, not well supported. First, feral or deprived children vegetate in an inhuman environment, which has therefore severe consequences for the physiological, psychological, cognitive, and social development in general. It seems quite naive to assume that the dramatic impact of deprivation can be reversed or should not influence learning (including language) after the child has been rescued. Second, one cannot draw direct comparisons between a neuropsychological recovery process and a typical acquisition process in children. One might say that there is a sensitive period for recovery from neurological language disorders but at the same time it cannot be concluded that the same process applies for typically developing children. Neural structures (re)organize throughout the life cycle, and it is not surprising that during the formation of neurons and connectivity in infancy and early childhood irreversibility of disorders is most promising and gradually decreases the more neural circuits become wired. However, this genetically determined neural developmental process does not prevent neuroplasticity as neural recovery occurs throughout the life cycle. New neurons are continuously born throughout adulthood and are integrated in existing neural formation. If the assumption of a critical recovery period is true, aphasic patients would not be able to recover at all, or only with minimal success. However, the clinical reality shows the opposite: it just takes more time than in youth, but neural plasticity provides good recovery at any stage of the life-cycle if the cortical damage does not exceed a certain degree of severity (Heiss et al. 2003).

12.2 The Multilingual Brain

Less than 25 % of about 200 countries in the world recognize more than one language as their official language. Only about a dozen countries recognize more than two languages as official (e.g., India, Luxembourg, Nigeria). However, globally there are more bilingual or multilingual speakers/signers[1] in the world than mono-

[1] The expression to speak a language or the term speaker is used here in a metaphoric sense as of course we also include signers.

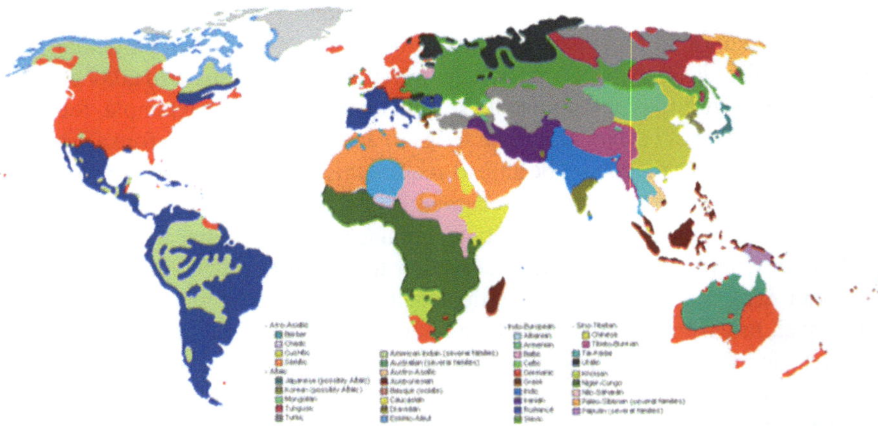

Fig. 12.2 Current state of language families. Often languages are grouped together such as the American (Indian) languages, although they are presumably not related. Much like biological organisms, languages diverge during the course of the history and often their common origin is not traceable. Minority languages are not considered. (Adapted and modified, Wikimedia)

lingual individuals as a result of formal education in the school or in the context of natural language converse. The main reason is that some languages are more frequently used for the purpose of global communication between different linguistic communities (e.g., English, French, Spanish, Portuguese, Arabic, Mandarin, Malay). But sometimes the usage of language is restricted to a geographically small region, which requires acquiring more than one language. An educated person in Eritrea, for instance, may learn to speak and write in Tigrigna, Arabic, and English—three native languages (L1s), which are quite different and which make use of the scripts Ge'ez, Arabic, and Roman (Fig. 12.2).

In principle, we can acquire or learn more than one language (L2+) at any time during our lifecycle. As with many other cognitive or motor skills, however, natural acquisition during childhood is to some extent more beneficial as neural circuits are in the process of being established; in contrast, the neural circuits of the adult brain are relatively fixed but neural plasticity allows modifying acquired neural circuits and/or creating new neural circuits. The child's brain develops new circuits without interference from already existing circuits and optimal cognitive shaping can be provided for a particular skill. The downside is that the child has no options, that is, it acquires what will be offered including less optimal input. For example, if the child automatizes Case marking errors because of its linguistic experience, relatively intensive formal training is required to correct these mistakes. In adults, however, often less well-established or automatized structures of a newly learned language are incorrectly produced as a result of interference with the dominant and automatized native language. Thereby, many factors influence the acquisition or learning process including the similarity between L1 and L2, age of acquisition, linguistic context, frequency of usage, and motivation.

Some basic L2 research questions are: To what extent does our brain processes nonnative languages differently from native languages? Does a possible processing difference between L1 and L2 depend on the degree of structural similarity and/or on a certain stage of brain growth? Is the number of languages our brain can handle limited or what are the benefits and/or the downside of speaking/signing more than one language? Although we will address these and other questions in the following, our focus here is on reviewing and discussing the temporal parameters and spatial locations and connections involved in L2 as compared to L1.

Before the introduction of electrophysiological and neuroimaging techniques in the 1980s and 1990s, observations and analyses of impaired L2 language in neurological patients served as main source for drawing conclusions about how the bilingual brain operates in comparison to the monolingual brain. As mentioned before, more than half of the world population can be considered as multilingual and thus that patients suffering from L2 language disorders is not an exception, but represents the majority of cases (Paradis 1998). The systematic diagnosis of L2 disorders in aphasia started with the use of the Bilingual Aphasia Test (BAT; Paradis et al. 1987). Specific psychometric and linguistic criterions were set for adapting the English version to other languages. Beyond standard neuropsychological test batteries, researchers evaluated language disorders in a customized fashion by presenting test material in a paper-and-pencil (offline) format. Thus, this neurolinguistic method described language disorders in relation to clinical symptoms and/or syndromes and tried to link these patterns to the lesion site assessed by CT scans.[2] It is apparent that this dual approach has its limits as it does not inform about the specific cortical regions, the neural circuits involved in L2 processing or about other cognitive functions, which may influence L2 processes such as verbal WM capacity, cognitive load, world knowledge, etc. Thus, to draw general conclusions about the language–brain relationship in bilingual or multilingual patients by observing and analyzing recovery processes is extremely difficult. However, some empirical findings are quite interesting.

In the study by Fabbro (2001), for instance, the recovery patterns of 20 right-handed bilingual Italian–Friulian aphasic patients, who acquired L2 between the ages of 5 and 7 years was as follows: about 65 % showed parallel recovery in both languages, 20 % were greater impaired in L2 and 15 % were greater impaired in L1. The interesting result is that there was no specific factor responsible for the recovery patterns. Neither the variables lesion type or site nor aphasic syndrome or pre-onset usage of L1 and L2 were responsible. In general, it can be therefore concluded that the combination of multiple factors seem to be responsible for an individual recovery process. Another finding refers to what is sometimes called pathological code switching (or language inference), that is, aphasic patients seem to suffer occasionally from impaired attention control to switch between both languages. For instance, the production of L2 words cannot be inhibited although the listener does not understand L2 (e.g., Mariën et al. 2005). Code switching disorders were associated with deep left frontal lesions. Here, we would need also to consider that the risk

[2] CT is an X-ray computed tomography.

of linguistic inferences between two languages is higher the similar both languages are. For example, one might expect more instances of inferences if the relevant language pair is Spanish and Italian rather than Spanish and Urdu. Let us look at two more examples. In the study by Fabbro (2001), agrammatic Italian/Friulian aphasic patients showed in general parallel recovery for both languages, but behaved differently with respect to pronoun omissions. This is not surprising when we take into account the typology of both languages. Italian is a pro-drop language (much like Spanish or Japanese for example), but not Friulian. Thus, if a pronoun will be dropped in Friulian, it is an obvious error, but this error cannot be detected in Italian as the pronoun omission is syntactically permitted. Similarly, English is a weakly inflected language, as it has no grammatical gender (but not in Old English); most Slavic and other languages have more than two grammatical genders; Romance languages typically use two different grammatical genders (female, male), but there are often exceptions and often linguists are required to account for specific morpho-syntactic patterns of a particular language. For instance, Spanish uses in addition to (fe)male markers, pronouns that do not have a gendered noun as antecedent but are neuter and refer to a whole idea, clause, or objects not mentioned in the discourse (e.g., *ello, esto, eso,* and *aquello*). The reader may realize that the observational method heavily relies on the behavioral–linguistic analysis while the associated neural correlates can be only broadly defined. It is desirable that the behavioral approach uses a typologically relevant analysis of the observed L2 patterns.

In this vein, it has been tried to link the behavior of outstanding personalities with exceptional skills to cortical properties that are different from the average person. In the domain of language, we refer here to the postmortem brain examination of the German sinologist/linguist Emil Krebs (1867–1930), who, according to family reports, mastered more than 68 languages verbally and in writing and had knowledge of about 120 languages. While there are good reasons to doubt that his language skills reached the online fluency level of 68 different native speakers, we can be certain that he was extremely polyglot. In other words, his metalinguistic knowledge and his ability of phonological modulation was exceptional good. Cytoarchitectonic differences between Krebs' brain and 11 control brains were analyzed with morphometry and multivariate statistical analysis (Amunts et al. 2004). The authors conclude that the Krebs' brain shows a local microstructural specialization (as compared to the control brains) for Broca's area: a unique combination of interhemispheric symmetry of BA 44 and asymmetry of BA 45 with respect to the right hemisphere.

These findings are difficult to interpret, as a unique exceptional brain cannot be compared. But let us assume for a moment that indeed a correlation between linguistic behavior and cortical structure exists in the case of Emil Krebs. Still, we cannot conclude that the cortical differences are actually related to linguistic computations *per se* or to cognitive operations supporting or providing the base for these computations. For instance, it is unclear whether cortical differences are related to high demands on WM functions, to operations associated with controlled switching between different languages (as required for translations), to the amount of lexical information processed or whether the results are coincidental unrelated to his linguistic behavior. However, in assuming that any highly repeated cognitive activity results in cytomorphological changes, much like people train their leg muscles to run faster, a correlation might be plausible in the case of Emil Krebs, but conclusions about neural correlates of a specific linguistic behavior remains highly speculative. Today, more direct methods are available to reveal the neural substrates of L2 processing. Let us turn therefore to electrophysiological and neuroimaging methods and studies that provide new insights about the neural correlates of bilingual processes.

Depending on a series of factors such as L2 proficiency, age of L2 acquisition, or structural (dis)similarities between L1 and L2, varying ERP/MEG findings have been reported. To begin with, the data reported do not support the account of a critical period of language acquisition (see below). A difference was found in late and early L2 learners by Weber-Fox and Neville (1996). While all groups (native speakers, and late and early L2 speakers) showed an N400 effect, they reported that late English L2 speakers (> 11 years of age) showed a delayed N400 of 20 ms compared to the other groups. In Hahne and Friederici's (2001) study, late L2 (L2: German; L1: Japanese) and monolinguals showed a similar N400 effect for semantically incorrect sentences. However, the N400 effect lasted ca. 400 ms longer in bilinguals than in monolinguals. The authors discuss the option that this delay may reflect the attempt of late L2 speakers to integrate the critical word in the sentence context as reduced lexical knowledge may prevent a fast decision comparable to the native speakers (see also Sanders and Neville 2003; Mueller 2005). Thus, the N400 effects found are quite similar among L1 and L2 speakers. The differences are mostly related to changes of latency and amplitude in late L2 speakers.

In the case of morphological complex words, Russian late L2 speakers of German showed a biphasic ERP waveform much like L1 speakers (Hahne et al. 2006). While incorrect participles elicited an early anterior negativity and a P600, incorrect plurals solely generated a P600. This finding is in line with the production proficiency level as the L2 speakers performed worse on plurals than on participles, probably due to differences in rule complexity. Thus, these data indicate that even late L2 speakers can reach native-like, automatic computations of morphologically complex words. A study by Rossiet al. 2006) revealed that age of acquisition is not necessarily the prime factor but proficiency. They found for late highly proficient L2 speakers of German or Italian and respective monolinguals comparable ERPs (ELAN, negativity, P600) for active voice sentences and agreement violations (LAN, P600). In contrast, minimally proficient L2 speakers elicited similar patterns for phrase structure violations, but only a P600 (not a LAN) for agreement violations. Moreover, the low-proficient L2 speakers showed a delayed P600 with reduced amplitude.

Fine-grained differences in syntactic L1 and L2 processing were reported in a series of MEG studies with Japanese (relatively) late English L2 learners (average age across studies: 25–28 years; Kubota et al. 2003, 2004, 2005). The first study tested Case violations checked phrase-internally (e.g., **I believe him is a spy*) or checked phrase-externally (e.g., **I believe he to be a spy*). Only the M150 (ELAN-like response at ca. 150 ms poststimulus) was reported for the phrase-internal checking violation in L1 speakers. L2 speakers seem not to be able to process this structure in an automatic fashion. The second study tested violations of noun phrase raising (e.g., **The man was believed (t) was killed*) and Case filter (e.g., **It was believed the man to have been killed*). Here, the Case filter violation did not elicit a M150 response, but the noun phrase raising violation did. Both L1 and L2 speakers showed this response pattern, indicating high-order syntactic sensitivity in L2 speakers. The third study examined infinitive (e.g., **He postponed to use it*) and gerund complement violations (e.g., **He happened using it*). Again, the gerund complement violation resulted in a M150 response for L1 and L2 speakers but the infinitive complement violations did not. Overall, these results show that only certain syntactic structures were processed in an automatic (online) fashion much like native speakers. Numerous MEG bilingual studies are published and refer to different linguistic levels (see for a review Schmidt and Roberts 2009).

Some fMRI studies were designed to find an answer for the basic question whether L1 and L2 would activate the same or different cortical regions according to age of acquisition. Kim et al. (1997) studied "early" (mean age 11.2 years) and "late" (mean age 19.2 years) bilingual speakers. The age of L2 acquisition was defined with respect to age when conversational fluency was reached in L2. The (healthy) participants were asked to silently generate sentences according to imagined events. The authors reported spatial differences in Broca's area in late bilinguals for processing L1 and L2, but early bilinguals activated for both languages common subregions of Broca's area. No differences were reported for Wernicke's region with respect to the age of L2 acquisition. Dehaene et al. (1997) reported sentence processing differences between L1 and L2 English–French speakers, whereas L2 recruited more right hemispheric activations. Only early bilinguals, who acquired both languages at birth, showed an overlap of activation for L1 and L2 (see also Perani et al. 1996; Saur et al. 2009). Two other studies did not reveal differences in a word-stem completion and sentence processing task performed by early (< 6 years of age) and late (> 12 years of age) bilinguals of Mandarin and English bilinguals (Chee et al. 1999a, b). However, the variable age of acquisition might not actually be the critical variable, at least at the level of sentence comprehension. Instead, the variable fluency (often to some extent interrelated to age of acquisition) seems to be important as highly fluent bilinguals activate similar left temporal lobe areas for L1 and L2, but not less fluent bilinguals (Perani et al. 1996). Very interesting findings stem also from a positron emission tomography (PET) study. PET scans were popular before the MRI technology became fully established. It is an imaging test that uses a small amount of radioactive substance (called a tracer). This neuroimaging technique has been superseded by MRI technology, although it is sometimes used to identify brain receptors (or transporters) associated with particular neurotransmitters. In this study, neural activity was measured during reading in German and English and translating words from German into English or inverse (Price 1999). L1 of the six participants was German and all acquired English as L2 at around 9 years of age. Compared to reading, the translation task activated cortical regions outside of the typical language areas, which involved the anterior cingulate and bilateral subcortical structures (putamen and head of the caudate nucleus).

Translation involves less automatized circuitries but a higher effort of coordination. In addition, control functions showed, during translation, higher activation of the supplementary motor cortex, cerebellum, and the left anterior insula. During language switching (not translation) an increase of activation was found in Broca's area and in the bilateral supramarginal gri. Thus, many neural activities related to processes between L1 and L2 occur outside of the typical language circuit. In another bilingual fMRI study it was examined how English L1 speakers process visually presented simple declarative sentences and signed sentences in comparison to signers of American Sign Language (ASL). The classical Broca–Wernicke circuit was activated in both languages, but in contrast to native English speakers, reliable activation was found in native signers (deaf or hearing) in posterior right hemisphere areas. This study confirms the particular role of the right hemisphere in visuospatial processing (Bavelier et al. 1998).

Let us look more closely at syntactic processing in bilinguals, a cognitive domain typically supported by Broca's region in L1 speakers. The fMRI study by Suh et al. (2007) revealed that for both languages (Korean: L1; English: L2) the left IFG and the (bilateral) inferior parietal gyri (among other areas) were activated when late bilinguals were asked to read center-embedded (e.g., *The director that the maid introduced ignored the farmer*) and conjoined sentences (e.g., *The maid introduced the director and ignored the farmer*). However, the left IFG (but not in any other areas) activity was higher for embedded vs. conjoined sentences in L1 but not for L2. The authors conclude that the same cortical areas are recruited for L1 and L2 syntax, but the underlying neural mechanisms would be different. These data are in contrast to the findings of Hasegawa et al. (2002), who reported that neural activity increased in L2 compared to L1 due to sentence complexity (negated vs. affirmative sentences). Suh and colleagues assume thus that in L1 less complex sentences may be processed in an automatic fashion while more complex sentences are not automatized and involve a higher cognitive demand. In L2, however, this difference cannot be detected as processing of different sentence structures would not have been automatized. This is a plausible interpretation. In the present case, syntactic complexity correlates with higher cognitive demands, but it is to assume that multiple linguistic and/or pragmatic information can be the source for increased neural activity.

A recent study, which used magnetic resonance DTI (see Basser et al. 2000), revealed white matter difference in L1 and L2 speakers (Mohades et al. 2012). The participants of this study were L1 speakers, simultaneous and sequential bilinguals (mean age: 9.5 years). Sequential refers to L2 acquisition >3 years of age, simultaneous to L1/L2 acquisition from birth onwards (L1 was either French or Dutch and L2 was a Romance or a Germanic language). One of the findings is that simultaneous bilinguals had higher mean fractional anisotropy (FA) values for the left inferior occipitofrontal fasciculus tracts (connects anterior regions of the frontal lobe with posterior regions in the temporal occipital lobe) than monolinguals. However, the comparisons for the fiber projection anterior corpus callosum to orbital lobe showed a lower mean FA value in simultaneous bilinguals as compared to monolinguals. In both cases, the sequential bilinguals had intermediate values compared to the other two groups. FA is here a measure for fiber density, axonal diameter, and myelination in white matter. It is therefore plausible to assume that the acquisition of two native languages at birth is beneficial for stronger and faster anterior–posterior fiber connections supporting language processing. However, since the myelination process of the fiber tracts is still ongoing in childhood, it might be that this outcome reflects only a particular time window of the white matter development. We cannot exclude the possibility that no significant FA differences will be measured for the anterior–posterior connection in adult monolinguals and bilinguals. If the fiber system is fully developed, a ceiling effect might have been reached. Thereby, we do not exclude the assumption of a lifetime learning process that can modify or change already established properties of the fiber connections. However, a postpuberty modification involve presumably different neural modifications than those in infantile brain development. The second interesting finding reported by Mohades and colleagues, lower mean FA value for simultaneous (early) bilinguals regarding the corpus callosum to orbital lobe connection, goes in line with the results that early bilinguals tend to be less left-sided lateralized for language than monolinguals or late bilinguals (Hull and Vaid 2006; Josse et al. 2008). Thereby, an increase in the size of the corpus callosum seems to correlate with a higher degree of left lateralization for language. These and other findings directly verify the assumption that the specific language acquisition process shapes the fiber system that is responsible for connecting different language relevant regions. In other words, cortical regions become language sensitive in a specific manner as the fiber system connects these regions according to the linguistic input received.

Some neurolinguistic findings show that late L2 speakers activate different cortical areas for L1 and L2. In contrast, there is clearly a tendency that early L2 speakers recruit the same cortical areas for L1 and L2. This general outcome is difficult to interpret: Do early L2 speakers rely on a single language system in opposite to late L2 speakers, who have different computational systems for L1 and L2? How many different language systems are then cortically represented in a different way in nonearly polyglot speakers? We do not have access to sufficient specific data to draw more general conclusions. L2 speakers vary in proficiency and fluency, and they use languages with different degrees of similarity, have experiences with different communication styles and domains, etc. Thus, it is not surprising to assume that every individual brain organizes language(s) in a different way.

Certainly, our daily observations tell us that young children acquire cognitive skills in a playful manner as compared to adults, whose learning process is apparently more effortful. However, does this imply that adults cannot reach the fluency or proficiency of a second language as young children do? The answer must be strictly denied. Everyone at any age can reach L1 fluency level in L2. Our brain is not an organ, whose functionality declines with the onset of adulthood. Brain plasticity and adult neurogenesis is a dynamic process and facilitates the acquisition of L2 proficiency in adulthood. Many variables would need to be considered to explain, why an individual acquires L2 knowledge in a specific manner. In general, it needs to be considered that it is difficult to capture neural activities requiring similar processing resources in L1 and L2. As pointed out before, given that morphosyntactic and phonological rules are different among languages, comparable structures in L1 and L2 may recruit different cognitive demands because of different degrees of automatized processes. However, L1 and L2 grammatical processes were supported in late bilingual twins by the same neural regions. The twins' (13 years of age) native language is Japanese, but they were trained during a period of 2 months on English verb conjugations. Pre- and posttraining fMRI study revealed increased activity in the left dorsal IFG, which correlated with their behavioral performance. Despite of significant proficiency differences in L1 and L2 with respect to the verb generation of past tense, the same cortical region was activated (Sakai et al. 2004). Similarly, when grammatical rules in a nonnatural, foreign language, which are inconsistent with the rules of natural languages, are examined, only the language-consistent rules activate Broca's area (Tettamanti 2002; Musso et al. 2003). This is confirmed by a recent fMRI study showing neural convergence in highly proficient bilinguals with respect to sentence comprehension and verb/noun production tasks (Consonni et al. 2013). Taken together, anatomical studies support the following view: If the L2 proficiency level matches native-level proficiency, common neural activities can be found in the left frontotemporal language circuit, but if the L2 proficiency level is clearly lower compared to L1, additional cortical activities are involved in the prefrontal cortex.

References

Amunts, K., Schleicher, A., & Zilles, K. (2004). Outstanding language competence and cytoarchitecture in Broca's speech region. *Brain and Language, 89*(2), 346–353.

Basser, P. J., Pajevic, S., Pierpaoli, C., Duda, J., & Aldroubi, A. (2000). In vivo fiber tractography using DT-MRI data. *Magnetic Resonance in Medicine, 44*(4), 625–632.

Bavelier, D., Corina, D., Jezzard, P., Clark, V., Karni, A., Lalwani, A., & et al.(1998). Hemispheric specialization for English and ASL: Left invariance-right variability. *Neuroreport, 9*(7), 1537–1542.

Chee, M. W. L., Tan, E. W. L., & Thiel, T. (1999a). Mandarin and English single word processing studied with functional magnetic resonance imaging. *Journal of Neuroscience, 19*(8), 3050–3056.

Chee, M. W., Caplan, D., Soon, C. S., Sriram, N., Tan, E. W., Thiel, T., & Weekes, B. (1999b). Processing of visually presented sentences in Mandarin and English studied with fMRI. *Neuron, 23*(1), 127–137.

Consonni, M., Cafiero, R., Marin, D., Tettamanti, M., Iadanza, A., Fabbro, F., & Perani, D. (2013). Neural convergence for language comprehension and grammatical class production in highly proficient bilinguals is independent of age of acquisition. *Cortex, 49*(5), 1252–1258.

DeCasper, A. J., & Fifer, W. P. (1980). Of human bonding: Newborns prefer their mothers' voices. *Science, 208*(4448), 1174–1176.

DeCasper, A. J., & Prescott, P. A. (1984). Human newborns' perception of male voices: Preference, discrimination, and reinforcing value. *Developmental Psychobiology, 17*(5), 481–491.

Dehaene, S., Dupoux, E., Mehler, J., Cohen, L., Paulesu, E., & et al. (1997). Anatomical variability in the cortical representation of first and second language. *Neuroreport, 8*(17), 3809–3815.

Fabbro, F. (2001). The bilingual brain: Bilingual aphasia. *Brain and Language, 79*(2), 201–210.

Greenough, W. T., Black, J. E., & Wallace, C. S. (1987). Experience and brain development. *Child Development, 58*(3), 539–559.

Hahne, A., & Friederici, A. D. (2001). Processing a second language: Late learners' comprehension mechanisms as revealed by event-related brain potentials. *Bilingualism: Language and Cognition, 4*(2), 123–141.

Hahne, A., Mueller, J. L., & Clahsen, H. (2006). Morphological processing in a second language: Behavioral and event-related brain potential evidence for storage and decomposition. *Journal of Cognitive Neuroscience, 18*(1), 121–134.

Hasegawa, M., Carpenter, P. A., & Just, M. A. (2002). An fMRI study of bilingual sentence comprehension and workload. *NeuroImage, 15*(3), 647–660.

Heiss, W. D., Thiel, A., Kessler, J., & Herholz, K. (2003). Disturbance and recovery of language function: Correlates in PET activation studies. *NeuroImage, 20,* 42–49.

Hull, R., & Vaid, J. (2006). Laterality and language experience. *Laterality, 11*(5), 436–464.

Josse, G., Seghier, M. L., Kherif, F., & Price, C. J. (2008). Explaining function with anatomy: Language lateralization and corpus callosum size. *Journal of Neuroscience, 28*(52), 14132–14139.

Kim, K. H. S., Relkin, N. R., Lee, K.-M., & Hirsch, J. (1997). Distinct cortical areas associated with native and second languages. *Nature, 388*(6638), 171–174.

Kisilevsky, B. S., Hains, S. M. J., Brown, C. A., Lee, C. T., Cowperthwaite, B., Stutzman, S. S., Swansburg, M. L., Lee, K., Xie, X., Huang, H., Ye, H. H., Zhang, K., & Wang, Z. (2009). Fetal sensitivity to properties of maternal speech and language. *Infant Behavior and Development, 32*(1), 59–71.

Kisilevsky, B. S., Hains, S. M. J., Lee, K., Xie, X., Huang, H., Ye, H. H., Zhang, K., & Wang, Z. (2003). Effects of experience on fetal voice recognition. *Psychological Science, 14*(3), 220–224.

Kubota, M., Ferrari, P., & Roberts, T. P. L. (2003). Magnetoencephalography detection of early syntactic processing in humans: Comparison between L1 speakers and L2 learners of English. *Neuroscience Letters, 353*(2), 107–110.

Kubota, M., Ferrari, P., & Roberts, T. P. L. (2004). Human neuronal encoding of English syntactic violations as revealed by both L1 and L2 speakers. *Neuroscience Letters, 368*(2), 235–240.

Kubota, M., Inouchi, M., Ferrari, P., & Roberts, T. P. L. (2005). Human magnetoencephalographic evidence of early syntactic responses to c-selection violations of English infinitives and gerunds by L1 and L2 speakers. *Neuroscience Letters, 384*(3), 300–304.

Lenneberg, E. H. (1967). *Biological foundations of language*. New York: Wiley.

Mariën, P., Abutalebi, J., Engelborghs, S., & De Deyn, P. P. (2005). Pathophysiology of language switching and mixing in an early bilingual child with subcortical aphasia. *Neurocase, 11*(6), 385–398.

Mohades, S. G., Struys, E., Van Schuerbeek, P., Mondt, K., Van De Craen, P., & Luypaert, R. (2012). DTI reveals structural differences in white matter tracts between bilingual and monolingual children. *Brain Research, 1435,* 72–80.

Moon, C., Cooper, R. P., & Fifer, W. P. (1993). Two-day-olds prefer their native language. *Infant Behavior and Development, 16*(4), 495–500.

Mueller, J. L. (2005). Electrophysiological correlates of second language processing. *Second Language Research, 21*(2), 152–174.

Musso, M., Moro, A., Glauche, V., Rijntjes, M., Reichenbach, J., Büchel, C., & Weiller, C. (2003). Broca's area and the language instinct. *Nature Neuroscience, 6*(7), 774–781.

Paradis, M. (1998). Language and communication in multilinguals. In B. Stemmer & H. A. Whitaker (Eds.), *Handbook of Neurolinguistics*. San Diego: Academic Press.

Paradis, M., Libben, G., & Hummel, K. (1987). *The bilingual aphasia test.* Hillsdale: Larence Erlbaum.

Perani, D., Dehaene, S., Grassi, F., Cohen, L., Cappa, S. F., Dupoux, E., Fazio, F., & Mehler, J. (1996). Brain processing of native and foreign languages. *Neuroreport, 7*(15–17), 2439–2444.

Perani, D., Saccuman, M. C., Scifo, P., Anwander, A., Spada, D., Baldoli, C., Poloniato, A., Lohmann, G., & Friederici, A. D. (2011). Neural language networks at birth. *Proceedings of the National Academy of Science, 108*(45), 18566.

Pinker, S. (1994). *The language instinct*. New York: Harper Perennial.

Price, C. I., Green, D. W., & von Studnitz, R. (1999). A functional imaging study of translation and language switching. *Brain*, 122, 2221–2235.

Rossi, S., Gugler, M. F., Friederici, A. D., & Hahne, A. (2006). The impact of proficiency on syntactic second-language processing of German and Italian: Evidence from event-related potentials. *Journal of Cognitive Neuroscience, 18*(12), 2030–2048.

Rymer, A. (1994). *Genie: A scientific tragedy*. New York: Harper Perennial.

Sakai, K. L., Miura, K., Narafu, N., & Muraishi, Y. (2004). Correlated functional changes of the prefrontal cortex in twins induced by classroom education of second language. *Cereb Cortex, 14,* 1233–1239.

Sanders, L. D., & Neville, H. J. (2003). An ERP study of continuous speech processing. *Cognitive Brain Research, 15*(3), 214–227.

Saur, D., Baumgaertner, A., Moehring, A., Büchel, C., Bonnesen, M., Rose, M., Musso, M., & Meisel, J. M. (2009). Word order processing in the bilingual brain. *Neuropsychologia, 47*(1), 158–168.

Schmidt, G. L., & Roberts, T. P. L. (2009). Second language research using magnetoencephalography: A review. *Second Language Research, 25*(1), 135–166.

Suh, S., Yoon, H. W., Lee, S., Chung, J.-Y., Cho, Z.-H., & Park, H. (2007). Effects of syntactic complexity in L1 and L2; An fMRI study of Korean–English bilinguals. *Brain Research, 1136,* 178–189.

Tettamanti, M., Alkadhi, H., Moro, A., Perani, D., Kollias, S., & Weniger, D. (2002). Neural correlates for the acquisition of natural language syntax. *NeuroImage, 17,* 700–709.

Weber-Fox, C. M., & Neville, H. J. (1996). Maturational Constraints on Functional Specializations for Language Processing: ERP and Behavioral Evidence in Bilingual Speakers. *Journal of Cognitive Neuroscience, 8*(3), 231–256.

Werker, J. F., & Tees, R. C. (1984). Cross-language speech perception: Evidence for perceptual reorganization during the first year of life. *Infant Behavior and Development, 7*(1), 49–63.

Prospects

The nature of language is already a discipline, which is part of neuroscience and neurobiology, even though not necessarily implemented in educational programs. It would be not wise to argue in favor of particular approaches as all endeavors furthering directly or indirectly our knowledge in this field, in particular negative evidence. We can however assume that certain approaches appear to be more suitable for exploring the interaction between neurobiological, cognitive and behavioral parameters. The science of language is as any natural science dualistic as universal concepts and instances are two sides of the same coin.

Taking a platonistic stance: If it is true that some chemical processes can be reduced to physical processes, does this unification process supports the idea of reductionism? Certainly not, unification makes use of concepts and refers to some instances. In other words, we would not be able to unify without dualism. The ability to develop and work with concepts is fundamental for humans. It is symbolic communication *per se*, including mathematics, which represents the source for our ability to create concepts. Concepts are partially based on instances (much like in natural languages) and are required for predictions and to recognize further relevant instances. The platonistic view applies to any science trying to unify concepts and instances. The concepts of the science of language cannot be reduced to concepts of natural sciences. The advancement, if we accept the idea of advancement for a moment, is that new technologies create new instances to help developing predictable concepts, and that new concepts help to develop new technologies. It is this interplay, which seems to drive the gain of knowledge. However, not all concepts can be empirically verified or falsified, but they do not need to be abandoned as they may be of great importance for new directions. What are the innovative directions in the near future, which may provide new insights about the human language system? Here, are some approaches, which appear to be promising:

- Mapping research on language and other cognitive-emotive domains
- Applying molecular neuroimaging to the domain of language processing
- Developing further neuronal net simulations
- Integrating cross-species research on communication

D. Hillert, *The Nature of Language*, DOI 10.1007/978-1-4939-0609-3,

As neural nets may share the same mechanisms across different cognitive domains, it seems to be important to unify theoretical frameworks across different cognitive domains such as language and music. Moreover, there is still a great need for integrating cognitive-emotive aspects and linguistic computations, which are inseparable at the empirical level. The development of more sophisticated neuroimaging technologies progresses and molecular neuroimaging seems to be the next promising step for researching the neural substrates of language processing. At the same time, progress has been and will be made on neural net simulations making predictions about linguistic computations, how they develop or break down in various clinical populations. Finally, it is important to integrate theoretical as well as empirical research across different species by considering our evolutionary path and speciation of other organisms. Only in the light of evolution, we are able to understand the human nature. More important, new research directions might arise due to developments in other fields not directly related to neurobiological or language-related issues. Although, disciplines become more and more specialized, they complement each other. It is desirable that the nature of language evolves to the discipline language sciences with further specialized subdisciplines. Although some scientific developments may lead to undesirable applications serving political purposes, basic science as here discussed has the purpose of knowledge gain and to improve or cure medical conditions. In the end it is language, which makes all this possible.

Index

D. Hillert, *The Nature of Language*, DOI 10.1007/978-1-4939-0609-3,

MIX
Papier aus verantwortungsvollen Quellen
Paper from responsible sources
FSC® C105338

If you have any concerns about our products,
you can contact us on
ProductSafety@springernature.com

In case Publisher is established outside the EU,
the EU authorized representative is:
Springer Nature Customer Service Center GmbH
Europaplatz 3, 69115 Heidelberg, Germany

Printed by Libri Plureos GmbH
in Hamburg, Germany